Alaska

Literary Field Guide

ART | ECOLOGY | POETRY

edited by
Marybeth Holleman, Nancy Lord,
and Shaelene Grace Moler

 See also Permissions, page 326.

Published by Skipstone, an imprint of Mountaineers Books—an independent, nonprofit publisher
Skipstone and its colophon are registered trademarks of The Mountaineers organization.

Printed in China

29 28 27 26 1 2 3 4 5

Design and layout: Jen Grable
Cartographer: Bart Wright, Lohnes + Wright
Front cover illustration: *Salmonbeary* by Kesler Woodward; back cover illustration: *Caribou* by "Chromatic" Crystal Jackson; spine: *Walrus oh-my-heart* by Alvin Amason
Frontispiece: *Ooh I keep coming back to you* by Alvin Amason
Illustration on page 6: *Marjorie Glacier* by Courtenay Birdsall Clifford; page 336: *Bears of Katmai* by Ray Troll
Map page 10, source: Alaska Native Language Center and the Institute of Social and Economic Research, University of Alaska, Anchorage

Library of Congress Cataloging-in-Publication data is on file for this title at https://lccn.loc.gov/2025026240.

Printed on FSC®-certified materials

ISBN (paperback): 978-1-68051-762-0

Skipstone books may be purchased for corporate, educational, or other promotional sales, and our authors are available for a wide range of events. For information on special discounts or booking an author, contact our customer service at 800.553.4453 or mbooks@mountaineersbooks.org.

Skipstone
1001 SW Klickitat Way
Suite 201
Seattle, Washington 98134
206.223.6303
www.skipstonebooks.org
www.mountaineersbooks.org

LIVE LIFE. MAKE RIPPLES.

Praise for Alaska Literary Field Guide

"The bountiful abundance of this wild collection is artfully packed for a bush plane flight through head and heart. Feel your way through this guide—and learn to love Alaska as much as I do."

—J. Drew Lanham, author of *The Home Place* and *Sparrow Envy*

"*Alaska Literary Field Guide* has described the indescribable: through poems, prose, and art, the sprawling beauty of Alaska settles onto the page like a woodstove's warmth on a frigid fall morning. Brew some coffee, get a copy, settle in."

—Hank Lentfer, author of *Raven's Witness* and *Faith of Cranes*

"Like Alaska, this book contains multitudes. Keep it by your desk as a tool for both reference and inspiration. By turns coolly scientific and ardently rhapsodic, these sharply focused portraits of Alaska animals, plants, and landforms keep us attentive to what is timeless and what is urgently at risk in our time."

—Tom Kizzia, author of *Pilgrim's Wilderness* and *Cold Mountain Path*

"Many ways of knowing coalesce in this essential collection. *Alaska Literary Field Guide* provides a full-spectrum, interrelated engagement of head and heart that truly honors the marvel of the living place we call Alaska."

—Derek Sheffield, Washington State Poet Laureate and co-author, *Cascadia Field Guide: Art, Ecology, Poetry*

keep coming back

To the gentle giant, the Steller's Sea Cow (*Hydrodamalis gigas*), and to the world's largest cormorant, the Spectacled Cormorant (*Phalacrocorax perspicillatus*), both of whom were hunted to extinction by whalers, fur traders, fox farmers, and other human commercial enterprises that failed to see you as more than meat, skin, and feathers. Our world is less for your loss.

May we do better by those who remain; may we honor their inalienable right to exist and all the myriad ways—only some of which we comprehend—that they make the world more vibrant, diverse, habitable, and beautiful.

Contents

Temperate Rainforest & Coastal Mountains 109

Boreal Forest 163

Rivers, Lakes & Wetlands 217

Tundra 261

Alaska

RUSSIA
ALASKA
CANADA
Attu Island
Bering Sea
Pacific Ocean
Utqiagvik
Arctic Ocean
Teshekpuk Lake
Port Hope
Colville River
Chukchi Sea
Brooks Range
Kotzebue
Seward Peninsula
Koyukuk River
Fort Yukon
NUNAVUT
Yukon River
Nome
St. Lawrence Island
Norton Sound
Fairbanks
Yukon River
Tanana River
YUKON
St. Matthew Island
Kuskokwim River
Denali
Alaska Range
Wrangell Mountains
Copper R.
Nunivak Island
Bethel
Chugach Mountains
Anchorage
CANADA
Bering Sea
Lake Clark
Kenai Lake
Prince William Sound
Kenai Peninsula
Mount St. Elias
Pribilof Islands
Juneau
BRITISH COLUMBIA
Bristol Bay
Alaska Peninsula
Kodiak Island
Gulf of Alaska
Sitka
Unimak Island
Aleutian Islands
Ketchikan
Unalaska Island
0
200
400 Miles
Pacific Ocean

Indigenous Groups of Alaska

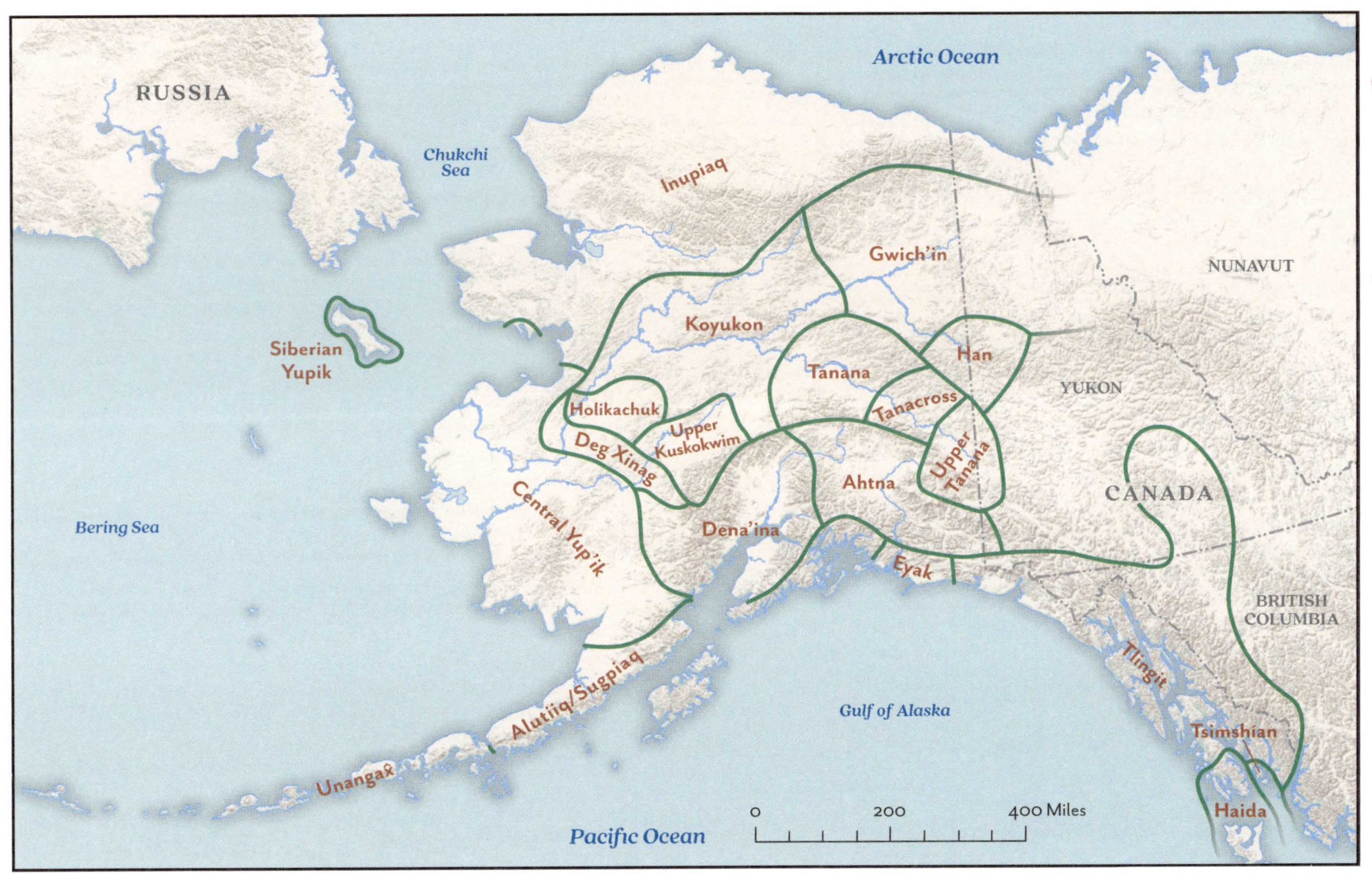

Introduction:

The Immense Wilds

This book is a love song to a place we humans call Alaska, from the Unangan word *alaxsxaq,* which means "the place where the sea breaks its back" or "great land." In the Unangax̂ known world, that land was what we refer to as the Alaska Peninsula—a small part of what is now named Alaska.

It *is* a great land. Is there any place in the United States more closely associated with stunning natural beauty, iconic wildlife species, and vibrant Indigenous cultures? Encompassing remote island shores crowded with fur seals and tide-water glaciers calving into the sea, the highest peak in North America and the vast wetlands of the Arctic Refuge, long-legged moose ranging in deep boreal forests and Dall sheep clambering up alpine ridges, Alaska is a land of wild beings and vast wilderness. Within these mostly intact ecosystems, Indigenous Peoples—including the Tlingit of Southeast, the Yup'ik of the western coasts, and the Athabaskan of the Interior—retain vibrant cultures wedded to land and sea.

The 730,000 people who call Alaska home enjoy such wealth and diversity year-round, and millions of visitors are drawn annually from around the world to our national public lands and communities. With this anthology, we provide a glimpse of what attracts visitors, and what enchants and holds us Alaskans here. Through poetry, art, and cultural and natural history notes, this book sings the praises of this great land, some of its amazing beings, and our relationships with one another.

KINSHIP PATTERNS, AND THE WEAVE OF ART + ECOLOGY + POETRY

Many traditional field guides related to Alaska already exist, but our approach has been different. Inspired by literary field guides such as *Cascadia Field Guide*, *The Sonoran Desert: A Literary Field Guide*, and *A Literary Field Guide to Southern Appalachia*, we've chosen to represent just a sampling of iconic or especially interesting beings. We've placed these into kinship patterns—that is, grouped them in the habitats where they relate. And we've combined Alaska's twenty different ecoregions into six major habitat types, with the understanding that most beings are found in many parts of the state and may range into habitats other than the primary one indicated here. Fireweed, for example, is grouped here with other beings occupying the boreal forest, though the plant grows from sea level to the edge of alpine tundra. Similarly, brown bears thrive along coastlines, in forested areas, and on tundra, where they're sometimes called barren ground grizzlies, but we've placed them in the temperate rainforest and coastal mountains habitat group.

For each featured being, we've written factual notes about biology, culture, history, relationships, even some curiosities and mysteries, and added poetry

and visual art by a diverse collection of contributors. This enables us to see each being through multiple complementary lenses, which together can reach us on a deeper sensory and emotional level. Poetry, with its interplay of wide-open form and evocative language, can allow the ineffable to be expressed, unveiling connections. Art, by forgoing words altogether and communicating through color and form, provides a more personal and interior interpretation that can transcend linguistic barriers. These diverse ways of knowing weave together into a tapestry of awareness that reveals the larger whole. Most important, this multidisciplinary approach recognizes that everything in our world connects to everything else, and that ecological understanding and environmental and cultural health are based on this interdependence.

WHO'S HIGHLIGHTED, AND WHY

Choosing a manageable number of representative beings to include was beyond challenging, as this place is immense in every way. The 663,000 square miles of Alaska's land contain thousands of kinds of beings inhabiting a diversity of terrain—fourteen major mountain ranges, vast sweeps of tundra, broad river deltas, and coastal icefields from which more than twenty-seven thousand glaciers pour. Alaska hosts more than one hundred species of mammals, six hundred species of fish, five hundred species of birds, and six thousand species of lichens and fungi alone. Alaska's oceans—the Gulf of Alaska and the Bering Sea of the North Pacific, and the Chukchi and Beaufort Seas of the Arctic Ocean—cover more than twice the total land area of the state and support more marine life than the rest of the nation's marine waters combined.

Given the thousands of lives at interplay in such a diversity of habitats, we aimed for a mix of iconic beings like salmon and polar bears

A Note on Naming and Kinship

We adopted from other recent literary field guides the use of the word *being* instead of *species*, the avoidance of referring to a being as an *it*, and the capitalization of beings' names within each ecological profile. These semantic choices, which honor the lives of other-than-human beings, were influenced by and are respectful of Indigenous worldviews that see all life as of equal value. To consider that everything has agency and spirit, has an inalienable right to exist, and deserves respect is to place ourselves, humans, as kin to all the rest. We are coequal and together, not above and apart. Such recognition can strengthen relationships and influence how we behave in and care for our world and all the beings who share it.

The Indigenous Peoples of Alaska, who make up about 20 percent of the state's population and speak twenty distinct languages, are especially close to and knowledgeable about the natural world, and we've tried to honor their experiences and values in both the creative work and the notes. When appropriate, we use the more specific names by which the people know themselves: Tlingit, Haida, and Tsimshian in Southeast Alaska; Eyak and Alutiiq/Sugpiaq in Southcentral; Unangan/Aleut in the Aleutians; Yup'ik in western Alaska; Athabascan, including Dena'ina and Gwich'in, in the Interior; and Iñupiat in the north.

with lesser-known beings such as the sea butterfly and the Steller sea lion, as well as some with particularly fascinating life histories, like our wood frog. We also strove for a balance among plants, mammals, birds, and other categories—paying attention to them all, from the large to the small. Sometimes, when we discovered a poem that exceptionally embodied a relationship or opened to fresh meaning, we took our lead from the poet. We went a step further, including among our selected beings some who in scientific terms wouldn't qualify as species but who are active community members in Alaskan culture, such as braided rivers, pingos, and glaciers.

SELECTING CONTRIBUTORS

Alaska is so large that to say one is from here belies the fact that most of us know only a small portion of our home state. Still, we wanted this volume to reflect knowledge acquired from long and deep connection to this place. We three editors are not cheechakos (newcomers): Nancy and Marybeth have lived in Alaska our entire adult lives, and Shaelene, born and raised in Kake, comes from generations of Tlingit in Southeast Alaska.

To begin the project, the three of us sought Alaska poets whose work we admired and invited them to contribute writing related to particular beings. Some of these pieces have been previously published and some were newly written for the book. We strove to include contributors who are diverse in every way—in style, region, ethnicity, gender, age, and background. Alaska is home to many, many accomplished poets and writers well versed in the life of their dwelling places, and we regret that we couldn't include more of them. We sought the same diversity from a smaller number of Alaska's outstanding artists, with a mix of existing and newly created art.

As we gathered poems and did our research, we became aware of two consistent threads. First, many Alaskans depend on the land and sea for food and livelihoods. Alaskans may be harvesters, hunters, and fishermen as well as observers, and we know other beings through direct and personal encounters. The selected poems frequently speak of such experiences in practical ways. We tried to reflect this reality in the profiles as well. We did not attempt to discuss management issues, as those are beyond the scope of this book. We also did not attempt to retell Indigenous stories or histories as they were not ours to tell, nor could we do them justice. While it was important to us that Indigenous perspectives were known and recognized throughout the text, we realize there may be shortcomings and sincerely apologize for these. We did our best to be inclusive without sharing any information that wasn't already public in order to respect Indigenous peoples' intellectual property.

A second thread relates to Alaska's rapid environmental change, including that related to a heating climate. Alaska is warming two to four times faster than the rest of the country, and the changes are evident everywhere. It's starkly clear that the beings of Alaska—human, salmon, caribou, black spruce, and the rest—are living in a time of great upheaval.

We thank everyone who's participated in our project and hope that this book will provide reading pleasure, engaging information, and inspiration to all who encounter it, whether from an easy chair by a window or beside a campfire beneath one of Alaska's magnificent mountains. This love song to Alaska is merely a sampling of all the fascinating and diverse wild beings who live here, something to whet appetites for discovering more of Alaska's endless wonders.

—Marybeth Holleman, Nancy Lord, and Shaelene Grace Moler

Marine Waters

With 47,000 miles of shoreline—more than the rest of the United States combined—Alaska is surrounded on three sides by rich and diverse ocean realms. The state's 200-mile-wide offshore Exclusive Economic Zone (EEZ) covers nearly 1.5 million square miles—more than twice the land area of the state—and is larger than the EEZ of all the contiguous forty-eight states combined. The North Pacific and Arctic Oceans include the Gulf of Alaska and the Bering, Chukchi, and Beaufort Seas.

These seas support the most abundant populations of fish, shellfish, seabirds, and marine mammals in the nation, and some of the most abundant in the world—among them tiny algae and copepods, cold water corals, hundreds of fish species, millions of seabirds, and marine mammals that include polar bears, walruses, seven species of seals, and fourteen species of whales.

There's good reason that so many beings either live in Alaskan waters or arrive in summer to feed. Three-fourths of the nation's continental shelf is off Alaska, and these shallow, cold, storm-tossed, nutrient- and oxygen-rich seas are among the most productive in the world. As well, several major ocean currents collide in Alaska's seas, among them the Alaska Coastal Current, the Kuroshio Current, the Bering Slope Current, and the Beaufort Gyre, all adding to ocean productivity. The abundance is also enhanced by wind-driven upwellings, which bring cold water rich in nutrients to the surface.

Humans, too, depend on these oceans. For thousands of years Indigenous Peoples co-evolved with both open water and sea ice, depending on both for sustenance, cultural vitality, and travel. Marine waters and ice coverage connect communities for visiting family and friends, hunting and fishing, and other crucial activities. Today, the largest wild salmon runs in the world pulse through Alaska waters, and half of all seafood commercially caught in the US is taken from them.

But our oceans, as vast and productive as they have been, are in trouble. Most of Alaska's threatened and endangered species are marine animals, and many seabird and marine mammal populations throughout Alaska are in decline—the result, scientists suspect, of excessive harvests of fish and marine mammals, habitat damage from bottom trawling, persistent organic pollutants, and long-term changes in the ocean environment. The world's oceans have been absorbing heat from the atmosphere—more than 90 percent of Earth's excess heat since 1955, when modern record keeping began. Warming oceans cause sea-level rise, coral bleaching, accelerated melting of tidewater glaciers and sea ice, intensified storms, and declines in ocean health and biochemistry.

A North Pacific marine heat wave in 2014–16 resulted in a mass of warm water (dubbed "the Blob" by scientists) that caused huge die-offs of seabirds, harmful algal blooms, starving whales, and fishery disasters. A layer of warmer-than-average water down to 300 meters persisted under nearly windless conditions for months, blocking the mixture of nutrients upward and affecting the entire North Pacific web of life. Sea surface temperatures were as much as 7 degrees F above average. Another marine heat wave in 2018–19 caused ten billion snow crabs to disappear from the Bering Sea; scientists have determined that they starved to death when rising temperatures required more metabolic energy and the crabs, confined to a smaller "cold pool," were unable to find enough food. These disastrous marine heat waves are expected to intensify in coming years.

Perhaps even more troubling, in recent decades the Bering Sea and Arctic Ocean have lost more than half of their sea ice, with severe consequences for the entire ecosystem. Sea ice is essential habitat for many marine beings, and as it disappears, shipping has increased, posing additional risks of ship strikes on marine mammals, underwater noise, air pollution, incursions by invasive species, and oil spills. With shrinking sea ice, offshore oil drilling and mining can more easily take place—and increase threats to the environment.

Another effect related to warming, ocean acidification, is the result of excess carbon dioxide being absorbed by the oceans. The acidity of seawater has increased about 25 percent from preindustrial times. This change in ocean chemistry makes it more difficult for calcifying organisms to build and maintain their shells. Of all the oceans, cold polar regions absorb the most carbon dioxide.

All of these remarkable changes in ocean health are already having serious impacts on Indigenous cultures, particularly those of the Iñupiat and the Yup'ik. Alaska's seas and coasts are unique and globally significant for their diversity, expanse, and abundance of fish and wildlife, and their historical, cultural, and economic importance. All life began in the oceans, but they are surprisingly underexplored and very poorly protected. It's clear that ocean health is essential for life everywhere on Earth.

Humpback Whale

Megaptera novaeangliae

Perhaps the most dramatic and iconic of all whales likely to be seen in Alaska's oceans is the Humpback, who often leaps from the ocean's surface in elegant acrobatic displays. Breaching (throwing themselves out of the water and landing with huge splashes), spy hopping (lifting their heads vertically), and slapping their flippers and tails on the surface are all part of the show. We don't know why they do this, but it's believed to be a way of communicating. Or maybe it helps remove pests from their skin. Or maybe it's just for fun!

This very large baleen whale—up to 50 feet long and weighing 40 tons—has long pectoral flippers that are white underneath, and tubercles (bumps) on the head and front edge of the flippers. Variations in black-and-white patterns and scars on the underside of the fluke (tail) make it possible for scientists to identify and monitor individual whales over time.

The common name of this whale comes from the distinctive hump on the back. In the whale's Latin name, *Megaptera* means "big-winged" (for those long pectoral flippers) and *novaeangliae* means "New England" (for the location where European whalers first encountered them). Humpbacks are known as Kun in Haida, the name for all baleen whales.

Humpbacks feed primarily on shrimplike zooplankton known as krill and on small schooling fish like herring. Each Humpback can eat up to a ton of food each day. Pleated grooves in their throats expand to allow Humpbacks to take in huge volumes of water and food; water is then forced out, catching the food in baleen (fringed plates in the mouth) so that it can be swallowed. Humpbacks also sometimes feed cooperatively in a method known as bubble netting, where a group of whales dives beneath a school of fish, blows bubbles to surround the school, then swims up together through the center of the bubble net to the surface, opening their giant mouths in unison to capture their meal.

Humpbacks are known for their elaborate underwater vocalizations. Both sexes and the young use calls to communicate, but only males are associated with long, haunting songs with complex, language-like structures. These choruses, which can be repeated for hours, change over time as individual whales embellish them. Like human languages, songs within a particular population are distinct from those of other populations but are passed around, sometimes across long distances, to influence others. No one knows for sure the purpose

of Humpback songs despite a great deal of study. One theory is that they act like birdsongs to attract mates.

Humpbacks inhabit oceans all around the world and travel great distances every year in what is one of the longest migrations of any mammal. Every spring, thousands of Humpbacks travel north from three different wintering grounds—Hawaii, the Western North Pacific, and Mexico—to spend summer feeding in Alaska's nutrient-rich waters. They are most likely to gather in Southeast Alaska, Glacier Bay, and Prince William Sound, although they're present farther west and north, and in recent years have been sighted more frequently in the Arctic. Recently some Humpbacks have remained in Alaska year-round, perhaps a sign of too little available food and the need to keep eating. Normally, the whales fast and live off fat reserves in winter.

Historically, Alaska's Indigenous Peoples harvested Humpbacks for food, oil, and materials for tools—and their baleen has been an important material for cultural expression. As it did most other whale species, commercial whaling in the 1800s and early 1900s devastated Humpbacks, reducing their population worldwide by 95 percent and in Alaskan waters to perhaps fourteen hundred individuals. Since receiving protections, they've increased to a peak of about twenty-one thousand in 2013. Formerly listed as endangered under the federal Endangered Species Act, the Humpbacks who migrate between Alaska and Hawaii—the largest of the three distinct populations—have been delisted since 2016. The populations who migrate to Mexico and the Western North Pacific remain listed as endangered.

Today the major threats to Humpbacks are ship strikes, entanglement in fishing gear, harassment or stress caused by whale watching, and climate change that can affect environmental cues and, most important, available food. The Pacific marine heat wave of 2014–16 resulted in an ocean hot spot nicknamed "the Blob" that caused a major decline in cold-water fish and plankton, which negatively affected both survival and reproduction of Humpbacks. The number of Humpbacks in and around Glacier Bay dropped by nearly 50 percent, and in Prince William Sound by 90 percent. None of these missing whales—known by their unique fluke markings—were found to have relocated to other regions.

Recovery for these large, long-lived mammals is very slow. Females—if they are well fed—give birth to a single calf only every two to three years. When there's not enough food, they don't reproduce. When they do, females nurse their calves for up to a year. If they survive to adulthood, healthy Humpbacks can, like humans, live very long lives—up to eighty or ninety years.

MARILYN SIGMAN

If the Light Is Right, and You Have Eyes to See

It's good practice looking for whales
to soften your gaze over the roiled surface.
Back and forth, back and forth,
watching in stillness for the change of the surface
when the whale reaches upward.
Ghost of water-drenched whale breath hangs like a flag,
a giant arrow pointing downward to the track of the whale.

So if your eyes are there, in that place, at that moment,
you're only one breath of the whale behind.

Whale follows plankton. Plankton follows
the great turning of salty mineral soup
brought by currents that track the centuries
of fiery basalt and rock torrents
tumbling down the slopes of continents
and dissolving into soup and plankton.

So, if our eyes are there, in that place, whale emerges against sky.
We're only one long breath of the planet behind.

Pacific Herring

Clupea pallasii

Each spring, after the long, dark winter, the waters of Sitka Sound come alive in an explosion of biological activity. Huge schools of Pacific Herring flood in from their winter habitat in deeper waters, turning coastal waters milky white with their spawn. Eagles and gulls swoop and dive, sea lions and harbor seals chase Herring, and humpbacks rise by the dozen to exhale their misty breaths.

The size of adult Pacific Herring—usually around 9 inches long—belies their incredible role in their ocean world. A keystone species, they anchor the marine and coastal food web. More than forty species—including humpback whales, bald eagles, brown bears, rockfish, and tufted puffins—depend on Herring as food at this critical time of year. As such, Herring are an important link in the marine ecosystem, transferring energy from zooplankton to higher trophic levels.

Like salmon, Herring exhibit "countershading" as camouflage from predators: they're dark blue-green on top, so that from above they blend with the darker waters, and they're bright silver beneath to blend in with the light from above. All together, they'll suddenly turn, flashing bright from below, further confusing predators.

During spawning, males and females release their milt and eggs into the water column, where they mix and fertilize. Then, the fertilized, sticky eggs (roe) attach to eelgrass, kelp, *Fucus* (known to Alaskans as rockweed or popweed), and other nearshore aquatic beings. While a female may lay twenty thousand eggs in one spawn, only two generally survive to become young fish. Eggs hatch in about two weeks. Herring larvae drift as plankton, and then juveniles mature in bays and inlets. The timing of Herring spawning evolved to coincide with the spring plankton bloom in coastal waters, providing larval and juvenile Herring with enough food to survive their first precarious months of life.

After spawning, adults return to offshore waters. There, they remain in deeper waters during the day and rise to surface waters at night to feed on zooplankton, small fish, and crustaceans. From maturity at age three or four, they return to nearshore waters each spring to spawn, through lifespans of about eight years in Southeast Alaska and twice that in the Bering Sea.

In Southeast Alaska, Indigenous Peoples continue the ancestral tradition of gathering egg-covered kelp and hemlock boughs (which have been cut and placed in the water as substrate) and then eating the eggs directly off the branch or preparing the eggs many different ways. For the Tlingit in particular, Yaaw (Herring) and their gáax'w (eggs) are tied not just to nutrition but also to identities and cultural practices. One anthropologist, who called this "a singular cultural phenomenon," found that 87 percent of harvested eggs change hands as many as four times as the circles of sharing extend throughout and beyond Alaska.

Herring populations fluctuate tremendously year by year, depending largely on environmental factors. Rising ocean temperatures disrupt the blooming cycles of plankton, the Pacific Herring's food, and can increase the mortality of young fish. Herring are also threatened by disease, changing ocean

conditions, and predator pressures. The Sitka fishery closed entirely for a couple of years when the fish were too small for commercial interests, only to rebound a year later with the largest commercial spring roe harvest in Alaska's history. The Prince William Sound population crashed in 1993, four years after the *Exxon Valdez* oil spill, and has not recovered.

The references to Herring that are found all over Alaska maps in many languages attest to this being's historical significance—and signal a warning. In Kachemak Bay, the town of Seldovia is named for the Russian word for herring. But overfishing ended that fishery. In recent years, a smaller number of Pacific Herring have entered Kachemak Bay in the fall, bringing with them feeding humpback whales.

These small fish have big lives: they travel from wide-open healthy oceans to small, quiet bays and lagoons, always together. Gathering by the millions, they dart and dive and turn and live as one.

VIVIAN YÉILK' MORK

A Herring Egg Love Poem That Pops in Your Mouth

A five-gallon bucket filled with herring eggs on hemlock branches
is delivered to my porch. If you receive herring eggs,
you know you're loved. Gunalchéesh.

Blanch eggs & eat them with soy sauce.
If you receive herring eggs, you know you're loved. Gunalchéesh.

I take herring eggs to friends & elders.
Gunalchéesh. If you receive herring eggs, you know you're loved.

This ritual, this food web, this tradition, is in danger of being lost.

Herring spawn areas, gone: Kah Shakes/Cat Island,
West Behm Canal, Ernest Sound, Hobart Bay, gone.
Seymour Canal, Chatham Strait, Hoonah Sound,
Tenakee Inlet, gone. Auke Bay, Lynn Canal, Icy Strait
and Yakutat Bay. Gone. Gone.

Tlingit testimony: 10,000 years. Our elders tell us life here
is not possible without herring.

If you receive herring eggs, you know you're loved.

All life feeds on herring & we share eggs with neighbors
down in Ketchikan & all the way up to Utqiagvik.

Gunalchéesh, you are loved.

Herring eggs fed our ancestors, herring are beaded
on our robes, woven through our stories & carved in masks,
& etched on silver bracelets.

The herring dances with us & mourns with us.
Herring: a celebration of our lives.

Gunalchéesh, if you receive herring eggs, you know you're loved.

Who will we be when there are no more herring?
Who will feed us & who will feed the creatures?

The herring web connects us like threads
of a woven robe. We want to wear this herring robe

for ten thousand more years. So, Dear Friend,
if you're invited to the Yaaw Ku.éex', the Herring Potlatch,

come celebrate with us—If you receive herring eggs there,
you know you're loved.

Fork-tailed Storm-Petrel

Hydrobates furcatus

Petrel is derived from *Peter*, the name of the biblical figure who walked on water, for petrels seem to do the same, pattering their feet over the surface of the sea, skipping lightly across some of the roughest waters of the world. Their genus name, *Hydrobates*, means "water walker."

These small seabirds are in the "tubenose" order of birds, the Procellariiformes, making them kin to albatrosses and shearwaters. The two tubes atop their beaks were once thought to be for excreting salt but are now considered a way of channeling airborne scents, both to sniff out plankton blooms miles away in the open ocean and to find their nests.

Alaska is home to two of the world's twenty species of storm-petrel: Leach's and Fork-tailed. The 2-ounce, 8-inch-long Fork-tailed Storm-Petrel is distinguished from most other storm-petrels, who exhibit darker shades, by a lovely

silvery-blue color that appears to change with the light. These northernmost storm-petrels navigate harsh winter storms with ease, zigzagging through wave troughs, then hovering and dipping to feed on zooplankton and small fish. Like many seabirds, they are long-lived: the oldest recorded was at least twenty-five years old.

They nest on islands offshore Southeast and Southcentral Alaska and the Aleutians, with the core of their population in the southern Bering Sea. Sometimes using old puffin burrows, they share these bird islands with thousands of other seabirds, including cormorants, puffins, murrelets, pigeon guillemots, murres, auklets, and gulls. These bird islands buzz with activity in summer; on Saint Lazaria Island in Southeast Alaska, half a million seabirds crowd onto 65 acres.

Fork-tailed Storm-Petrels are among the most difficult seabirds to observe: they spend most of their lives over the cold, remote waters of the open North Pacific Ocean and come ashore only for nesting and feeding their chicks, even then arriving only at night and not during stormy weather. During such weather, chicks can survive several days by reducing their body temperature and entering a state of torpor, pausing growth. When seas are particularly intense, the adult birds sometimes seek refuge on the decks of commercial fishing boats.

Fork-tailed Storm-Petrels have been seen diving into water around surfacing baleen whales to nab plankton and fish escaping the whales' maw. Bering Sea Indigenous Peoples call them O ku ik, "bird that eats oil," because they gather around carcasses of marine mammals to consume the fat-rich floating oil. This they regurgitate to feed their young; they also regurgitate it onto predators and in times of stress. They coo and twitter to their nestlings at night, but at sea they remain silent. To spy one dancing with massive ocean swells is like beholding a mirage, a glimpse of the vastness of life the open oceans hold.

SARA LOEWEN

East Amatuli

eight small pieces of whale fat, some feathers, the lens of a fish or squid's eye, five smoothly worn bits of pumice

—from the stomach of one fork-tailed storm petrel

Summer nights in the Barrens,
northernmost islands in the archipelago,
are raucous with hundreds of thousands of petrels
swooping like winged fragments of nimbostratus
over steep green slopes.

Because petrels are philopatric
—home loving—returning to the same burrows and same mates
with whom they share the work of tending to a single chick,
the baby a soft downy ball, lavender-gray,
a powder puff musky sweet as motor oil,
I want to write them into pleasant metaphors.

Because they're moon-guided,
numinous and undaunted in storms,
I want to praise the way these birds, so clumsy on land,
are nimble on the wing.

Buoyant ramblers that skim and patter,
hovering on wave-made air
glancing from wave crests to wind shadows,
arriving as a sailor's omen, a portent of gales.

Leaving before sunrise, landing after sunset,
—darkness being their defense against ravens, eagles and gulls.
Only the brightest full moon keeps petrels from the sea
where, drawn to lighthouse beams and lights of ships,
they were caught by whalers and sealers
who threaded wicks through their lithe bodies
to draw up oil and burn as candles.

Sea Butterfly

Limacina helicina

Pteropods (from the Greek for "wing foot") are marine gastropods—free-swimming sea snails and sea slugs. What would be a foot in the rest of the snail world is divided into two parts and propels these creatures through the water just as the wings of an insect do in air. Pteropods fall into two categories—the shelled kind, Sea Butterflies like *Limacina helicina*, and the unshelled or "naked" kind commonly known as sea angels.

Sea Butterflies are found in all the oceans, but a person is unlikely to meet one in nature, as they're small (the size of a lentil), live below the surface of the water (though not too deep), and are easy to overlook. Once in captivity, they do very poorly. Their shells and wings are nearly transparent, allowing their soft parts (often colorful orange or purple) to be visible. They are, in fact, quite beautiful small creatures when viewed under a microscope, worthy of their comparison to butterflies.

Limacina helicina lives in cold northern waters—in the Arctic Ocean and the northern reaches of the Atlantic and Pacific Oceans, where this being makes up a large proportion of all resident zooplankton and is considered to be a keystone species. Sea Butterflies feed on algae and smaller

planktonic species that they capture in mucus webs and then suck into their mouths. Sometimes called the potato chip of the sea, they, in turn, are fed on by a variety of fish, whales, seabirds, and larger zooplankton species. Studies have found that this pteropod makes up about half of the diet of juvenile pink salmon.

Sea Butterflies are at serious risk from ocean acidification. As the oceans absorb carbon dioxide from the air, helping to buffer climate change, seawater becomes more acidic. Because the thin shells of Sea Butterflies are composed of a kind of calcium carbonate, this acidification of their ocean home makes their shells weaker and can lead to complete loss of the shells. A study in nearshore waters of the Beaufort Sea found that 70 percent of the pteropods there are already suffering weakened shells. Future consequences are likely to be grave—not only for Sea Butterflies themselves, but also for the entire web of life they support, from the great whales to the smallest of fish.

NANCY LORD

"Pteropods make thinner shells"

The science paper says just that.
The latest research shows
the little snails are slimming down.

Shells like tiny cinnabuns
rolled tight and crisply brown
are stained and scanned to show the proof.

Light as toast or deeper brown.
Brighter is thicker, darker thinner,
with measurements in micrometers.

Our carbon-filled, fast-changing ocean
is sliding down the acid scale
to test our stressed-out kin.

More photos—shells like silver jewels,
coiled in nature's perfect design,
protective, enduring, since dinosaurs dwelt.

But now, cloudy here and pitted there.
The words are *dissolution, damage*.
Whorls weakened, flimsy, structure failing.

And the bodies, the soft, still-hidden bodies.
They can only do so much.

Sugar Kelp

Saccharina latissima (formerly *Laminaria saccharina*)

When the tide is high, they're an underwater jungle of dancing green, red, and brown fronds, sanctuaries for marine life, from the tiniest salmon smolt to the humpback whale. When the tide recedes, they lie in a tangle, harboring crabs and sea stars, keeping them wet and safe. More than five hundred species of seaweed grow along Alaska's coastline, most of whom are brown algae called kelps—the fastest-growing plants in the ocean.

Thriving in cold, nutrient-rich waters, kelps are foundation species who create the most extensive marine vegetated ecosystems in the world. Their thick stands—called forests—buffer coastlines from waves and erosion, reducing wave sizes by up to 60 percent. They provide safe habitat for a host of marine life: sea otters, seals, sea lions, and whales rest and feed in them; herring spawn on them; sponges and limpets grow on them; dozens of species of fish use them as nurseries; crustaceans, marine snails, sea urchins, sea stars, octopus, and fish call them home. Studies have counted more than one hundred thousand species living within mature kelp forests.

Newly discovered fossils show that kelps have been around for more than thirty-two million years and gave rise to dramatic biodiversity. Evidence also suggests that the presence of a nearly continuous "kelp highway" along the coastline influenced the migration of humans from Asia to the Americas. Kelps remain a foundation of Alaska's coastal Indigenous cultures.

Like all kelps, Sugar Kelp is not a true aquatic plant but rather an alga. Instead of roots, kelps use holdfasts to attach to rocks and other surfaces on the ocean floor. The holdfasts of Sugar Kelp are finely branched, with short and flexible stemlike stipes. The crinkly, ribbonlike blades grow up to 8 inches wide and 16 feet long. Sugar Kelp thrives to depths of 100 feet in protected subtidal and lower intertidal zones, but can, with surging storm waves, dislodge and wash to shore.

This is where most humans see them: along the high tideline in a tangled row called beach wrack. Here this and other seaweeds continue to provide habitat and food for springtails, flies, and shorebirds. Sugar Kelp is harvested by humans for both food and garden fertilizer. This kelp's common name comes from the sweet white powder called mannitol, which surfaces as the blade dries.

Kelps' absorption of carbon dioxide and production of oxygen is important for other beings and reduces the amount of carbon dioxide in our oceans, helping to limit ocean acidification. Sea otters in kelp forest ecosystems help keep grazers like sea urchins in check, allowing the forests to flourish. When sea otters were hunted to near extinction in the eighteenth and nineteenth centuries, sea urchin populations increased dramatically and toppled kelp forests, creating lifeless "urchin barrens." Even as numbers of sea otters in parts of Alaska's waters increase and contribute to kelp forest health, other factors related to climate change are causing kelp declines worldwide.

In recent years, kelp farming has become a fast-growing industry in Alaska. Sugar Kelp is one of the three main farmed species; the others are bull kelp and ribbon kelp. These farms involve lines suspended vertically in the water and require little tending to produce commercial crops.

Sugar Kelp is a generous being. Every part at every life stage—holding fast to the ocean floor, lying in mats at low tide, or resting in wrack along the shoreline—provides food and shelter to a rich marine community.

KERSTEN CHRISTIANSON

Carol's Laminaria

Luminary,
your flame draws you
to the rocky beach
where the kelp
washes upon the shore:

slick, brown blades coiled
by bull kelp tossed
with popweed.

You carry an armful
of slippery laminaria
to the tree line;
think of women hanging salmon,
while draping your haul
over a branch
to sway,
sing, and dry
in the wind.

In photographs
you capture the whorls
and turns of sea
tangle. You leave
behind your beach
finds: smooth stick,
broken shell. I look

for your signature,
your thumbprint,
your gathered kelp
this side of the equinox,
this side of the sea:
gold thread embroidered,
cupping the moisture
of drizzle and herring snow
in March.

Beluga Whale

Delphinapterus leucas

With their white backs arcing gracefully from Cook Inlet's gray waters like rolling whitecap waves, the smaller dark gray backs of their calves bouncing along beside them, Beluga Whales are aptly called Qunshi or Quyushi by the Dena'ina, which translates as "that which comes up." The Beluga's Latin name translates as "white dolphin without a wing," the wing in this case being a dorsal fin. The Beluga Whale has, instead, a dorsal ridge that allows navigation with ease through and under densely packed ice. The common name, Beluga, derives from the Russian word for white.

Belugas are small toothed whales who live throughout the Arctic and subarctic. In Alaska, they form five genetically and geographically distinct populations. The Bristol Bay, Eastern Bering Sea, Eastern Chukchi Sea, and Beaufort Sea populations are all protected under the Marine Mammal Protection Act. Only the Cook Inlet population, once thought to number in the thousands and estimated at only 331 in 2023, is further protected under the Endangered Species Act.

Alaskans and visitors delight in being able to spot Cook Inlet Belugas from downtown Anchorage and the highway along Turnagain Arm. These endangered Belugas were once found throughout the inlet but in recent years have concentrated in the upper inlet. Although they've been protected from hunting or harassment since 2005 and are the subject of a federal recovery plan, their numbers are not increasing. The reasons for this are unknown but may relate to some combination of threats—lack of prey, noise and other human disturbance, ship traffic, pollution, habitat loss, disease, and climate change.

When born, Belugas are dark gray and weigh about 100 pounds. They turn white as they mature and may live for seventy or more years. A large male can be 16 to 18 feet long and weigh more than 3,000 pounds, with blubber making up half of its weight.

Belugas have made their way into popular culture via Raffi's famous "Baby Beluga" song, various children's books and toys, and at least one Disney film. They're especially attractive because of their expressiveness—mouths that curve as though smiling and unfused neck vertebrae that allow them to turn their heads. Most people first (or only) encounter Belugas at marine parks and aquariums, although marine mammal capture from the wild is prohibited by most countries today and Beluga captivity is being phased out.

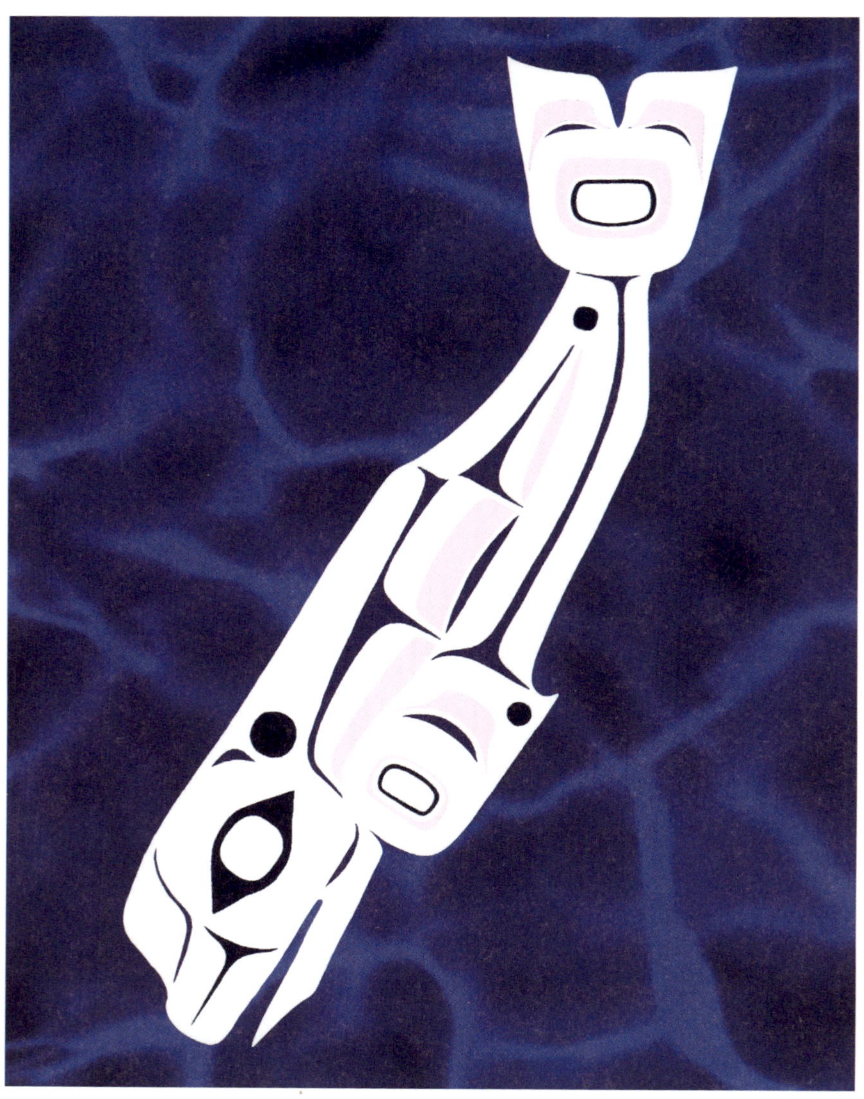

These small white "smiling" whales are thought to be highly intelligent and may have the most well-developed echolocation abilities of all marine mammals. They reshape their large melons—the bulbous tops of their heads—to focus and project sounds in the dark Arctic waters and in the shallow, muddy waters of Cook Inlet. The wide range of high-pitched, birdsong-like Beluga vocalizations, used for echolocation and communication, has led to the nickname "canary of the sea."

KIM CORNWALL

What Whales and Infants Know

A beluga rising
from the ocean's muddy depths
reshapes its head to make a sound
or take a breath.
I want to come
at air and light like this,
to make my heart
a white arc above the muck of certain days,
and from silence and strange air
send a song
to breach the surface
where what we most need
lives.

Pacific Halibut

Hippoglossus stenolepis

They begin life as tiny larvae, drifting with currents and feeding on phytoplankton. As they grow, they look like all the other small fish of the sea, swimming upright with one eye on each side of their heads. But when they're about six months old and an inch long, they metamorphose into a flat fish: their left eye migrates around to the right side of their head, and they descend to the ocean floor to begin their long lives as Pacific Halibut.

From larvae, Pacific Halibut grow to become the largest flatfish in the world. Older female Halibut can weigh more than 500 pounds and be 8 feet long and 5 feet wide. They grow so large because they scoop up everything in their paths and can live for fifty years.

Carnivorous, Pacific Halibut eat other fish, octopus, crabs, and clams. When Halibut are younger, other fish and marine mammals prey on them. But as they grow, Halibut become one of the dominant seafloor predators.

One of more than a dozen flatfish species in Alaska, Pacific Halibut range and migrate throughout Alaska's continental shelf, from Southeast to the Aleutians

to the Bering Strait. Many other flatfish species, like yellowfin sole, share their preference for shallow waters; Pacific Halibut may descend to about 1,500 feet when spawning, but otherwise remain in waters from 20 to 1,000 feet deep.

Male Halibut don't spawn until they're around eight years old, and females not until they're twelve. In groups at the edge of the continental shelf, they "broadcast spawn": several females and males release eggs and sperm into the water at the same time, increasing the chances that the eggs will become fertilized and decreasing the chances of predation.

Because of human fishing, few Halibut survive to live more than twenty-five years. Older and larger Halibut, especially females, are important for population stability, as female Halibut grow more prolific with age; while a 50-pound female produces about half a million eggs, a 250-pound female can produce four million eggs.

Since the 1970s, Pacific Halibut have been getting smaller for their age. By the 2000s, an average twelve year old weighed only half as much as one in the 1980s. This size decrease may be resulting from a combination of competition for food, climate effects, and commercial fishing.

These smaller Pacific Halibut are also susceptible to pollution, overfishing, and bycatch in trawl fisheries. The biggest threat may be the deoxygenation and warming of seas wrought by climate change; researchers predict that suitable habitat for Pacific Halibut, particularly juveniles, may disappear altogether off the coasts of Oregon and Washington by 2100 and shrink by 50 percent in Alaskan waters.

Coastal Indigenous Peoples, from the Siberian Yupik of St. Lawrence Island to the Unangax̂ of the Pribilofs to the Tsimshian of Metlakatla, have long appreciated Pacific Halibut within their cultures. Every language group has its own name for this being, including Chagix̂ in Unangan and Ivisa in Siberian Yupik. For the Haida, Pacific Halibut are a symbol of prosperity, and their potlatch dishes are frequently carved in the shapes of Halibut. Some Indigenous Peoples, as they do with salmon, make a special offering of the first Halibut they catch each season.

Among the most powerful fish in the sea, these muscular swimmers glide, swimming sideways, along the ocean floor. When they stop, with their countershading of white below and mottled brown above, they blend in with their surroundings, disappearing.

LESLIE LEYLAND FIELDS

The Halibut

The halibut lies on the butchering table,
her morose, thick-lipped face hanging
off one edge and
both eyes looking at me.

She lies flat, the way she used to swim
with no sides, just top and bottom
and movement like a ripple.

Her back is a map of the ocean floor—
dappled with kelp-colored spots
and moss-green blotchings like the black sand
and sea grass she hides in.

I can see underneath, where she is white,
only white with just the trail
of her backbone down the center—
there's a terrible delicacy in that
and I cannot touch her there.

My father comes out now with a knife and asks,
Do I want to know how old she is?
Smiling, he digs his blade in a circle
beside the staring eye, fingers the hole
then offers to me on his fingertip—something
white. A chip of shell maybe,
and beautiful, like china.

Ear bone, he says.
You can count the rings like a tree
to see how old she is.
We lose count at thirty.

Orca

Orcinus orca

Every year the fjords of Southcentral Alaska host gatherings of more than 150 sleek, powerful, black-and-white whales, slapping side flippers, breaching, spy hopping, vocalizing, and lobbing their tails. This magnificent "superpod" aggregation is an annual gathering of resident Orca family groups, or "pods," important for socialization and breeding.

Orcas occupy all oceans but favor cold, highly productive seas, with Alaskan waters supporting one of the largest Orca populations in the world. Alaska's roughly two thousand Orcas are most often seen over the continental shelf and among the Aleutian Islands, where upwelling and currents create ecologically rich waters. As sea ice declines, Orcas are now also venturing into Arctic seas. With their black-and-white coloring, small white patches near their eyes, and the tall triangular dorsal fins of males, they're among the most distinctive and easily seen whales in Alaska.

Highly social and intelligent, these stunning whales—who are actually classified not as whales but as dolphins—have three distinct ecotypes, each with different diets, behaviors, and genetics. Residents form family groups of up to forty individuals and eat only fish, especially silver and king salmon. Transients, who travel more widely and in smaller groups, prey on marine mammals, primarily seals and porpoises. While these two groups' ranges overlap, they don't socialize or interbreed. A third ecotype, offshore Orcas, remain farther out to sea and are so rarely encountered that little is known about them.

The Orca's other common name, Killer Whale, comes from a mistranslation of what early Basque whalers called them: *asesina-ballenas*, meaning "whale killer." Some researchers still prefer to call them Killer Whales, since a pod of Orcas is capable of killing a large whale. While historically, cultures with little knowledge of Orcas feared them, there's no known incidence of a wild Orca killing a human; the only known human fatalities have been at aquariums with captive Orcas.

Alaska's Indigenous Peoples have long viewed Orcas more accurately: to them, Orcas symbolize family, longevity, harmony, community, and protection. They're a prominent feature of totem poles and coastal Indigenous stories. In Orca's Tlingit origin story, a human named Natsilane carved the first "black fish" from yellow cedar; he taught them to eat seals and halibut but not to harm humans. Instead, they protect humans and help travelers find their way home. The Haida see the Killer Whale people, Sgaana xaaydagaay, as the most powerful people from the sea. In Alaska, Orcas are also called Sea Wolves, from the Tlingit *Gonakadet*. And, indeed, Orca social organization is remarkably like that of wolves.

Orcas have a strong and complex social structure. Pods are matriarchal, composed of multigenerational groups; both males and females stay with their mother their entire lives, splitting off into subpods only if the pod becomes too large. Even then, subpods regularly visit. All hunting is cooperative, and Orcas' exceptionally developed echolocation allows them to generate three-dimensional auditory "pictures" so precise that they can pick a silver salmon out of a school of mixed fish.

Within these pods, Orcas have distinct vocalizations, so distinct that researchers can identify pods or even subpods by their language, as well as by their unique saddle-patch markings. Clicks are used for echolocation of prey, pulsed sounds are thought to be used for group recognition, and whistles are used for up-close socializing. Other marine mammals can also distinguish resident from transient Orcas: in Prince William Sound, the well-known AB pod adopted a Dall's porpoise, who remained with them for years. This Dall's porpoise would have been food for the AT transient pod, who also frequent the sound.

Long-lived and slow to reproduce, males generally have a life span of thirty to fifty years, and females can live well past that, sometimes into their eighties. As males mature, the crescent-shaped dorsal fin straightens and grows to up to 6 feet tall. Females mature at twelve to seventeen years old, and their calves, small and bouncing along beside them, nurse for a year or more.

As true apex ocean predators, Orcas have only one true predator—humans. In the 1980s, Orcas learned to snatch black cod from commercial longlines as fishermen were hauling the lines from the seafloor. Some fishermen illegally shot the whales and used underwater explosives to deter Orcas from their fishing lines. The problem subsided with greater public scrutiny, enforcement, and changes in fishing gear. In the 1960s, more than two hundred wild Orcas

were captured for display at marine parks, decimating some pods such as the Southern Residents.

What most threatens Orcas now are oil spills, contaminants, underwater noise, lack of prey due to competition from commercial fisheries, vessel strikes, and climate change. The genetically distinct AT1 pod, who—unlike most transients—remain loyal to western Prince William Sound, lost fifteen of their twenty-two members as a result of the 1989 *Exxon Valdez* oil spill. With no reproductive females, the unique AT1 pod is now expected to go extinct. The resident AB pod also suffered huge losses from the oil spill; their numbers subsequently began to grow but then declined after the 2014 to 2016 marine heat wave, reversing thirty years of post-spill recovery.

Intelligent, agile, and highly social, Orcas are extremely dependent on, and epitomize, healthy oceans. They interact with their ocean world in every way: beaches at just the right incline and cobble size are used as "rubbing beaches" where Orcas playfully ride the waves and rub their bellies on the rounded cobble, then slide gently back into the sea.

CARROL RICHARDS

There's a cry so deep it comes from the ocean.

There's a cry so deep it comes from the ocean.
They say that after an orca dies, her male offspring perish within a year or two.
My mother died on March 22, 2013; she was 85.
Female orcas typically live about 50 years but can reach 90 in the wild.

They say that after an orca dies, her male offspring perish within a year or two.
My brother Tommy died on March 31, 2014; he was 64.
Female orcas typically live about 50 years but can reach 90 in the wild.
The average lifespan of male orcas is 30, but they can live up to age 60.

My brother Tommy died on March 31, 2014; he was 64.
My brother BJ died on January 10, 2015; he was 62.
The average lifespan of male orcas is 30, but they can live up to age 60.
My brother BJ died at home in Kiana.

My brother BJ died on January 10, 2015; he was 62.
His plaid shirts hung in the open closet.

My brother BJ died at home in Kiana.
After he died, I slept in his bed.

The risk of death in older male orcas increases eightfold after
the loss of their mother.
My brothers followed my mother.
After he died, I slept in his bed.
There's a cry so deep it comes from the ocean.

Pacific Walrus

Odobenus rosmarus divergens

Soft, high-pitched moans and deep guttural grunts rise from the shore, and then a sound that cannot possibly be coming from such large, bulky animals: a high ringing, like strings plucked on a violin. This chiming comes from the Walrus, the last living species in the family Odobenidae, distantly related to seals and sea lions. The Walrus's distinctive features have long fueled stories and songs.

Walrus tusks, which can grow to more than 3 feet long, are elongated canine teeth that continue to enlarge throughout their lives. Employed for dominance and display, the tusks are primarily used to pull the Walrus out of water and onto sea ice—so it's no surprise that the scientific name *Odobenus* means "tooth walker." The stiff bristles surrounding their tusks are highly sensitive organs with which Walruses seek and find invertebrate prey in the seabed. (The plural of *Walrus* can also be *Walrus*, although that use is less common.)

Walruses' tough, wrinkled skin can be several inches thick and lightens from a dark brown to almost pink as they age. Insulated by a thick blubber layer, Walruses also control blood flow to their extremities to regulate their body temperature—which is why they sometimes appear reddish. And they have an air sac under their throats that allows them to bob vertically in water and sleep. Living up to forty years, a male Walrus can weigh more than 2 tons, while females

are about half that size. Males possess the largest penis bone of any animal; it is popularly known as an oosik, from the Iñupiaq word *usuk*.

Known as Aiviq in Iñupiaq, the Pacific Walrus (one of two subspecies of walrus, the other being Atlantic) plays important roles in Arctic Indigenous cultures. This being's meat, fat, hide, bones, and tusks have been essential to human survival along Alaska's coast, and thus the Pacific Walrus was key to the development of Bering Sea societies. The Pacific Walrus is often featured in Indigenous artwork and storytelling, often as a shape-shifter with humans.

Living along coastlines and on pack ice and ice floes, Alaska's Walruses migrate hundreds of miles seasonally, following advancing and receding sea ice. They spend summers north of the Bering Strait in the Beaufort and Chukchi Seas. In spring and fall, they drift with ice through the Bering Strait, to winter as far south as Bristol Bay.

Walruses feed on the shallow continental shelf, foraging in seabed sediment and using their snouts to dig up clams, snails, worms, sea cucumbers, and tunicates. By disturbing the seabed, they release nutrients into the water column, providing food for a host of other species.

Walruses depend on sea ice reaching over the continental shelf. Sea ice provides resting platforms over seabed feeding areas, and females require ice for giving birth and raising their pups, who remain on the ice while mothers search for food. With the ice pack shrinking and pulling away from shallow waters, females must range farther in search of food, increasing pup mortality. Walruses now crowd shorelines in summer, with as many as thirty thousand ashore in one place; this results in more stampeding deaths, especially among the young. And now with more ship traffic across an ice-free Arctic Ocean, Walruses are increasingly threatened with ship strikes and oil spills.

While orcas and polar bears sometimes kill young and infirm Pacific Walruses, their tusks, bulk, and strength make them formidable opponents. They were decimated by meat and ivory hunting during the days of commercial whaling, but such hunting has been banned since 1941. The Marine Mammal Act of 1972 protects them today from harassment and from hunting, with an exemption for subsistence use by Alaska's Indigenous Peoples.

As an iconic, keystone species in the Arctic, Pacific Walruses help thread that icy ecosystem together. They're also among the first to suffer when that ecosystem falters. As formidable as they appear, they are also vulnerable, and unique among all living mammals.

NANCY DESCHU

Shapeshifting Home

Thousands of male walrus
lie in heaps on a narrow beach
on this small island in the Bering Sea.
They fan with their flippers,
shade their cinnamon faces,
stroke their large bellies.

The females are swimming north
to the Chukchi Sea,
their hundred-pound young cling
to their backs, a long migration
to feed on mollusks and crustaceans,
to rest on ice floes.

On the island, a male awakes, barks—
lumbers over piles of walrus.
A fight breaks out, tusks joust and stab,
blood drips from wrinkled necks.
He enters the water, a graceful swimmer,
and sings a chiming sound
like a bell lightly rung.

Calm settles again on the beach.
But the calm is deceptive—
water warms, ice transforms,
habitat is shapeshifting.
The sea rises, beaches shrink,
the walrus fall out of their rest,
into the sea.

Sea Ice

When the sun slides below the horizon in Arctic winter, leading to months of darkness, and temperatures plummet to –30 degrees F, Sea Ice is quiet and still, ethereal in its vastness. In summer, when the sun circles the sky for months, Arctic Sea Ice pulls north, detaching from coastlines, breaking with sharp cracks and echoing groans. Floating in the farthest north reaches of Earth, Arctic Sea Ice harbors a surprising and little-understood diversity of life above, below, within, and along its edge.

Historically covering about 7 percent of Earth's surface, Sea Ice forms in both the Arctic and Antarctic but is far more extensive and remains year-round in the relatively enclosed Arctic Ocean. As winter approaches, Arctic Sea Ice expands southward and by March can reach as far south in the Bering Sea as Bristol Bay and the Alaska Peninsula.

Advancing and retreating, breaking apart and grinding together, Sea Ice forms when surface seawater freezes, at about 29 degrees F, beginning in calm waters as thin sheets called nilas or in choppy waters as disks called pancake ice.

Larger sheets collide into dramatic pressure ridges and hummocks, and spread apart to reveal narrow open water leads and wider, longer-lasting openings called polynyas. Shorefast ice holds to a coastline, some anchoring to the seabed, and pack ice drifts with ocean currents and winds, a mesmerizing show. Multiyear ice increasingly loses salt and hardens, growing up to 13 feet thick.

Life thrives on, below and along the margins of Sea Ice. Brine channels percolate down through the ice, harboring a unique community of microorganisms that complete their entire life cycles within or attached to the ice. Ice algae in particular play a critical role at the base of the polar food web. Many Arctic marine fish, birds, and mammals, including polar bears and all four of the ice seals—ribbon, spotted, bearded, and ringed—are entirely dependent upon Sea Ice for breeding, resting, denning, pupping, and feeding.

Melt along the retreating ice edge in springtime creates upwelling currents, bringing nutrients to the bottom of the ice, creating an ecologically critical ice-edge phytoplankton bloom. This bloom forms the base of one of the most direct food chains of any ecosystem on Earth: phytoplankton to zooplankton to arctic cod to ringed seals to polar bears.

Arctic Sea Ice reflects most sunlight—the "albedo effect"—keeping waters beneath it a constant cold temperature, winter and summer, affecting ocean circulation and regional weather, and moderating the climate of the entire northern hemisphere. The ice pack also dampens ocean waves in storms, protecting shorelines from erosion.

But as the climate warms, Sea Ice is forming later in the year and ice seasons are becoming shorter and less predictable. Coverage is now shrinking at a rate of more than 12 percent per decade, and over the past thirty years the oldest and thickest Arctic Sea Ice has declined by an astounding 95 percent. As the reflective ice surface shrinks, seawater is increasingly exposed to sunlight and warms even faster, accelerating the depletion of Sea Ice and increasing atmospheric warming. This positive feedback loop is why the Arctic is warming three or four times the rate of the rest of the planet—a phenomenon called Arctic amplification. And as evaporation increases from all this open water, rain-dominated precipitation increases across the Arctic, causing profound effects on tundra ecosystems.

These extreme changes to Sea Ice threaten the culture, identity, and livelihoods of Indigenous Peoples across Alaska, especially the Iñupiat and the Yup'ik. When the protection of coastlines by Sea Ice is lost and storms increase, erosion eats away at the land. Some low-lying villages are having to relocate while others struggle to survive within their subsistence cultures. While Arctic people have many words for different types of Sea Ice given the crucial role ice plays in their lives and cultures, some of these words, including the Yup'ik word *Tagneghneq* (thick, dark, weathered ice), are now becoming obsolete. In contrast, the Iñupiaq word *Sikulġauraq* may be growing in usage; it refers to young, thin ice that can't be trusted to support the weight of a whale killed for subsistence.

Scientists predict that the Arctic will lose nearly all its summer Sea Ice within the next decade. Such catastrophic disappearance of this shape-shifting floating world will be among the most extensive losses of ecological habitat on Earth.

EVA SAULITIS

The Ice of Norton Sound, Refusing

for Kristine

Like any fixation, any obsessive
collecting, it's steeped in a specialized

language. She tries to catch each
word on her tongue, taste each separate

fragment. She closes her eyes like one pretending
to know how to describe the aftertastes

of wines: abrasive—the frazil,
slithery—the grease, the hiccup

of shuga, this process she repeats, learning
the lingo, then studying the guides,

identifying the mammals, the plants.
But now there are no plants, just

shadows, no colors, just hues
of transparency, grays that haven't

been identified. So she keeps a life list
of phenomena: water sky, ice blink,

kapsik, pressure ridge, avalanche
of pack ice barreling down

the Bering Strait to make of the sea
an anxious self. So she embarks

uninformed, terrified, without
a lexicon or pencil, strapping on

snow shoes, heading straight to where
a lead splits the frozen sea in two,

a place where knowing
ends, where language founders,

where wind carries off all her word
endings. Remember that edge,

ships plunging off explorers' maps
& someone's marked the margin: *Here there be*. . . ?

But the crucial part's worn away—?
Remember? That's where she begins.

No received notion of ice or self, no ecology
ties the jammed maze together.

She finds no edge, just the fractured remains
of ice that once held a shape, her shape, ice

of endless kinds, refusing to be named.

Red King Crab

Paralithodes camtschaticus

For years, viewers have been captivated by one of reality TV's most popular shows, *The Deadliest Catch*, which showcases all the drama of Bering Sea crabbers fighting high seas and winter weather to drop and then haul back pot after pot filled with giant, squirming king crabs. The real stars should perhaps not be the crabbers on the boats but the amazing animals who spend their lives on the bottom of the Bering Sea.

Alaska's Red King Crab is one of forty species of king crabs in the world, and one of three that are commercially harvested in the state. These red-colored beings are also known as Kamchatka Crabs. They live in cold northern waters, not only in the Bering Sea but also along Alaska's southern coast and all the way out the Aleutian chain and to Russia.

They are enormous—weighing up to 24 pounds with leg spans that can stretch to 5 feet. Males are larger than females, and the two can be told apart by the shapes of their "tails" or abdominal flaps—narrow on males and wide on females. These crabs sport five pairs of legs: the first pair with claws, the next three for walking, and the last, smaller set for specialized use—in males for transferring sperm to females during mating and in females for cleaning their fertilized eggs.

Females brood thousands of fertilized eggs beneath their abdominal flaps for an entire year. The eggs hatch into swimming larvae to feed on plankton, go through several molts (acts of shedding shells), and eventually settle on the bottom as baby crabs the size of dimes. A crab's shell is an exoskeleton, and crabs keep molting as they grow; new shells build inside the old ones and eventually split them open to become the new, roomier shells.

Adult crabs migrate annually between shallow and deep waters. They molt, mate, and hatch their eggs in shallow waters in late winter, then move to deeper (colder) waters during warmer periods, where they generally separate into groups by sex. Adult males have been known to migrate for distances of 100 miles annually, sometimes moving as much as a mile per day.

King Crabs are omnivorous and feed on invertebrates as well as any dead or decaying organic matter, plant or animal. In various stages of their lives they become prey, in turn, to fishes, octopuses, and some marine mammals. King Crabs' large size and delicious taste to humans make them the most coveted and monetarily valuable of crabs—and the most valuable fisheries species in the North Pacific, a true delicacy for those who can afford the high price. Families

and communities catch King Crabs for their own use, but those catches are dwarfed by the size of commercial catches.

The commercial fishery, however, is on the wane. The Bering Sea fishery was closed entirely for two years and opened for just a small take at the end of 2023. (The allowable harvest of just over 2 million pounds compares to the 1980 peak of nearly 130 million pounds.) Sharp population declines of Red King Crab are likely due to several factors. Sea ice is a critical component of King Crab survival; larvae depend on the early algal bloom that grows on the underside of the ice. Ocean warming limits the "cold pool" that persists at the bottom of the Bering Sea throughout the summer, ocean acidification can inhibit shell growth, and trawl nets used in other fisheries can damage bottom habitat. Other Alaska crab species, including snow crabs, have experienced a dramatic decline in the Bering Sea for similar reasons; nearly the entire population of Bering Sea snow crab died, most by starvation, after record-breaking ocean bottom temperatures in 2018 and 2019.

Deep under the surface waters, beyond human sight, Red King Crabs tiptoe over the ocean floor on long, prickly legs, in search of food in cold and colder waters. Protected by their sturdy shells, they mate, molt, guard their eggs, and play essential roles in seafloor foraging.

MARIE TOZIER

King Crab

Mighty crustacean
Of the Arctic sea
Baited into waiting pot,
And raised
From beneath thick ice
Only to be lowered in a boiling bath
And steamed
To bright perfection.
Do you feel pain?
Does it even matter?
Know that I will savor
The edible parts of you:
Your thick leg meat,
The portion hidden
In the tip of your claw
—If I were my grandmother,
The last sound would be
A long slurp as I drink and swallow
Broth from your shell.

Polar Bear

Ursus maritimus

Looming large in our collective imaginations and stories, walking the farthest reaches of land and Arctic sea ice, massive beings as white as the frozen stratum upon which they live stand not only at the top of the world but also at the top of the Arctic food chain.

The Polar Bear is both the largest of all bears and the world's largest land carnivore. Large males can weigh up to 1,700 pounds and stretch over 10 feet in length. Because Polar Bears live primarily on sea ice, where they hunt for seals, they're considered marine mammals—hence the scientific name, which translates to "maritime bear." The Iñupiat know this being as Nanuk, a word made familiar to English speakers by the 1922 Robert Flaherty silent film *Nanook of the North*. Of the nineteen Arctic subpopulations, two primary subpopulations live in Alaska—those of the Southern Beaufort Sea east of Utqiagvik, and those of the Chukchi Sea to the west.

Polar Bears are exquisitely adapted for survival on sea ice and in Arctic waters. They have large feet to distribute their weight when walking on snow and thin ice, and short ears and tails to conserve heat and avoid frostbite. They are superbly insulated with thick layers of fur and fat. The fur consists of a dense undercoat covered with long guard hairs that look white but are actually translucent and have a highly porous core that traps air and provides insulation. These bears are also strong swimmers; their body fat and hollow guard hairs help them float, and with their large front paws they can "dog paddle" for days and hundreds of miles.

Female Polar Bears den on either sea ice or land by excavating holes under snowdrifts or against a bank or bluff. Cubs (usually twins) are born in dens in December or January and weigh only a couple of pounds at birth. Polar Bear milk is extremely rich in fat, and cubs grow quickly. Mothers and cubs typically emerge from dens in early April, and cubs remain with their mother for two more years. Polar Bears are not true hibernators, and females who don't give birth and all males remain active all winter.

The Polar Bear has become the iconic poster animal of climate change because their sea ice home is literally melting from beneath them. With the loss of sea ice due to global heating, they're deprived of the habitat they need for all aspects of their lives, from hunting to denning to resting. Forced to spend more time on land, they struggle to find enough fat-rich foods. Although Polar

Bears are excellent swimmers, stormy seas and increasing distances between sea ice and land are not only taxing their fat reserves but also proving too much to survive; increasing numbers of Polar Bears, particularly females with cubs, have drowned. Because of the threat of extinction posed by the loss of sea ice habitat, in 2008 the Polar Bear was listed as threatened under the federal Endangered Species Act. While the status of the Chukchi population remains unknown, the Beaufort Sea population is in clear decline.

As Polar Bears are forced to spend more time on land, and as grizzly (or brown) bears move northward, the habitats of the two converge and some cross-species breeding has occurred. These rare hybrids are known as pizzlies or grolars, and in Iñupiaq as Nanulak. Polar Bears and grizzlies are genetically close, thought to have diverged some four hundred thousand years ago, during a glacial period.

Polar Bears have always had close relationships with the Indigenous Peoples of the North, who honor their strength, courage, and spiritual power. Alaska's Iñupiat understand this being's behavior and habitat through extensive generational knowledge. In recent years they've documented population and behavioral changes due to the loss of sea ice, contributing significantly to scientific knowledge and management decisions.

As sea ice shrinks and Polar Bears are stranded on land for longer periods of time, Polar Bear tourism has appeared in Alaska. In the fall, visitors fly to Iñupiaq villages like Kaktovik and Utqiaġvik, where Polar Bears scrounge the bone piles of subsistence-hunted bowhead whales while they wait for the pack ice to expand again toward the coast.

When the ice approaches shore, the great white bears waste no time heading for their offshore home, where they can resume their lives upon sea ice and seek their favorite fat-rich prey, ice seals. Then, finding a seal breathing hole, they'll crouch and wait, perfectly still in the Arctic silence, only the black of their noses visible in the vast expanses of ice.

DG NANOUK OKPIK

Whiteout Polar Bears

Dead on—in the night sky,
or stuck in the deep web,
bear stars exist. Name the
bone piles on the marsh
heaving like the Chukchi
sea-pure-ice-white and
Arctic air rising. Fifty miles
of open water floating.
I'm a carcass with
marrow bones 5x's an ice
bear at 1,500 lb. and 9 feet
tall. One swipe of my paw
you're neck-snapped, to the ice,
melt ground, cheek to red
ice stream. I glance across
the whiteness to myself.

Intertidal & Coastal Shores

Encompassing rocky headlands where sea lions sleep, curves of beaches where mergansers feed and river otters cavort, grass-lined lagoons where brown bears graze, and mudflats where thousands of shorebirds rest, along with islets and islands and spires and intricate passages and fjords, Alaska's shores are stunning in their beauty and diversity. And in their sheer size, with 6,600 miles of coastline—more than the rest of the nation combined.

When all islands and inlets are counted, Alaska has more than 47,000 miles of intertidal zone. It's a dynamic place, subject to significant uplift and subsidence from seismic events, glacial activity, permafrost thaw along the Arctic coast, intense storms, sea ice scouring, and large volumes of freshwater discharge from rivers.

As well, Alaska's twice-daily tides, in some places ranging as much as 30 feet in six hours, are among the largest in the world. This enormous tidal range arises from the shape of the coastline and bottom topography, together with the nearness of continents to each other in the northern hemisphere.

All these factors combine to create a rich diversity of habitats and a varied coastline replete with natural wonders like tidally reversing waterfalls, swaying kelp forests, and rock seawalls dripping at low tide with many-colored anemones, sea stars, encrusting corals, and mussels. These diverse intertidal and coastal habitats, some sheltered and some exposed to powerful waves, provide homes for abundant, unique, and resilient forms of life. The communities living "between the tides" are determined by the amount of sunlight, availability of nutrients, length of immersion in waters of variable salinity, air temperature, wave action, and interspecies interactions.

Rocky intertidal areas—such as those in Southeast, the western Gulf of Alaska, and the Aleutian Islands—may be either sheltered or exposed, and host biological communities of attached organisms such as algae, chitons, limpets, barnacles, mussels, and sponges; or slowly mobile organisms like grazing snails, sea stars, and sea urchins. *Fucus* (also known as rockweed, popweed, or bladderwrack), a brown algae with small air bladders suspending the plant in the water at high tide, grows abundantly on rocks in the midzone of most rocky intertidal areas.

In contrast, sediment-dominated intertidal areas—such as those along the Bering Sea and Arctic coast—host communities of invertebrates that live within and atop mobile sediments, such as mud and sand. Mudflats are important winter feeding areas for waterfowl and are vital stopover areas for migrating shorebirds; for example, up to 30 percent of a western sandpiper's diet during migration consists of the small macoma clam found there. And at high tidelines on beaches, beach wrack hosts a myriad of decomposers.

Distinct "biobands" form within the intertidal zone, depending on the water level within the zone. In the upper intertidal, biobands (from top to bottom) include biofilm such as the noticeable "bathtub ring" formed by mats of black lichen (*Verrucaria*) or blue-green algae, then barnacles, rockweed, green algae, blue mussels, and red algae. In the lower intertidal, biobands (from top to bottom) include red algae, alaria, brown kelps, surfgrass, and eelgrass. Each bioband has its own distinct color; shades of black, white, golden brown, emerald green, dark blue-gray, red, bright green, and dark brown present a glistening, kaleidoscopic rainbow of life.

Intertidal areas are critical resting and reproductive habitat for several marine mammal species, including fur seals, walruses, Steller sea lions, harbor seals, and sea otters. Terrestrial animals like minks, bears (brown, black, and polar), and deer forage along beaches. Many of Alaska's pink salmon and herring spawn within the intertidal zone. As well, many types of seabirds and sea ducks nest and feed close to shore and shelter in bays and coves, among them the colorful harlequin duck.

During low tides, tide pooling is a favorite recreational activity for Alaskans. Many beings remain still while exposed to air, waiting for the incoming tide, and when submerged become active once again. These include crabs, anemones, nudibranchs (sea slugs), octopuses, sea stars, and small fish like gunnels. Low tide is also the time for humans to pry edible foods like chitons from rocks and for other foraging. In Southeast Alaska alone, scientists have counted at least 170 species of intertidal invertebrates.

Onshore areas above the reach of tides are home to salt-resistant and hardy plants like beach rye, sedges, beach greens, beach peas, and beach sunflowers.

Alaska's intertidal areas, as vast and diverse as they are, are at risk from shoreline development, invasive species, pollution, and climate change. The 1989 *Exxon Valdez* oil spill caused serious damage to intertidal ecosystems, and oil persists even today in intertidal sediments. Marine debris, including lost fishing gear, plastics, and toxic materials, collects continuously along Alaska's intertidal zone, entangling animals and polluting habitat.

As the place where the worlds of sea and of land intermingle, Alaska's intertidal areas exert a powerful and radiating impact on the health of all of Alaska's habitats and beings.

Bald Eagle

Haliaeetus leucocephalus

The symbol of the United States of America, the Bald Eagle, is one of our country's conservation success stories. In the late twentieth century, due largely to the use of the pesticide DDT, this large bird of prey nearly disappeared from the Lower 48. After DDT was banned in 1972 and the Bald Eagle was listed as endangered under the Endangered Species Act in 1973, populations began recovering, and this being was removed from the list in 2007. Alaska played a role in this recovery—from 1982 to 1990, 279 Alaskan eaglets were relocated to the Lower 48.

Bald Eagles have always been held in great respect by America's Indigenous Peoples. Both ravens and eagles live at the heart of cultural beliefs and practices in Alaska, with Bald Eagles representing power, protection, and seeing the world with great vision. Both birds are well represented in Indigenous art, including house posts and totem poles, and Eagle feathers are used ceremonially. In Tlingit culture, Eagle is Ch'áak'. The Haida distinguish a juvenile Eagle (Gúud) from a mature, white-headed one (Ts'áak').

Despite their central role in Alaska's Native cultures, Alaska's Bald Eagles were not always thought of kindly. During the "bounty years" of 1917 to 1953, up to two dollars was paid by the territorial government for each pair of Eagle talons, because Eagles were seen as a threat to salmon fisheries. Although 128,000 pairs of talons were paid for during those years, Alaska's Eagles were never seriously depleted. Today, with protected status, they number approximately 150,000.

The Bald Eagle is not, like a vulture or condor, actually bald. The name refers to the white head feathers of the adult birds. (The scientific name in Latinized Greek means "sea eagle with a white head.") Mature birds have both white heads and white tails, while immature ones are a mottled brown and sport

dark rather than yellow bills. It takes about five years for a bird to transition to mature plumage. Immature Bald Eagles can be mistaken for the less abundant golden eagles, who generally live in more interior parts of the state.

Eagle pairs typically mate for life, and their courtship can sometimes be spectacular, involving locking talons in mid-air and tumbling nearly to the ground. Females are noticeably larger than their mates; in Alaska, they can weigh up to 15 pounds.

These birds nest in sturdy trees and, in treeless areas, on cliffs, usually in sight of water. Their nests, built of sticks with softer material interwoven, are very large—up to 8 feet across and a ton in weight. Both members of a pair collect materials, but the female does most of the construction. Nest preparation usually begins in February, followed by egg laying. Eggs (usually two) hatch in April or May and the young eaglets fledge in June or July. The same nests are usually used for many years, with yearly remodeling.

Although Bald Eagles are associated with eating fish, they also catch birds and small mammals (watch out for your cat or small dog!) and often feast on carrion and garbage. Given their wingspan and weight, they can lift up to 6.5 pounds of prey from a dead stop position, but with momentum they can lift much more. Flying in to catch fish, they've been clocked at more than 30 miles per hour.

Some of the best places to spot Bald Eagles, especially in winter, are around landfills. Where rivers remain open in winter, large convocations (as groups of eagles are appropriately called) gather to feed on late-spawning salmon. Near Haines, in the Chilkat Bald Eagle Preserve, between three and four thousand Bald Eagles gather during October and November. The Bald Eagle Foundation in Haines hosts a corresponding Alaska Bald Eagle Festival each year.

The Bald Eagle's descending whistling call is very distinctive but surprisingly high pitched for such a large and noble bird. In movies and on TV shows, listen for the common substitution of the red-tailed hawk's stronger, more resonant cry.

PEGGY SHUMAKER

Rapt

Glazed aspen, low sun, snag branches
snowed over except
where eagles choose to rest.
One jangles her head—neck

feathers splay, troubled and wrung.
She sneezes, then preens
tucking white over brown
lifts off, weightless,

catching the perfect surge, kicking
back a burst of white off laden
boughs, skimming
low over the vast

open river, the last in Alaska, wide Chilkat
running warm in December. A choir
of her sisters swivels
to watch.

Whoom, whoom, she's fast along the surface,
the inner curve of her wings filling
with light. Feet first, talons
flick-snatch a flap of silver

from the river. She lifts and banks,
someone *screes*. Not five feet
from me, she lands, spreading
her catch lengthwise

along the branch. Her claws
pin it still gasping—
her sharp beak
punctures. Liquid bites, *chlap, chlap*

so close I hear them
drip, hear the wet rip
as fish muscle turns
into eagle. What's not eaten

shivers, ripples. Slippery entrails spill,
shimmer over the limb. Iced
cottonwoods creak. Grainy wind
lifts sun-sparked snow. She glances

my way
without concern, her yellow
eyes intent. I shake
my head and feel

what has been wind-torn
mend.

Tideline

As the tide recedes, it leaves behind a tangled line of material. This Tideline, or wrack line, constantly changes and shifts; varying tides mark many Tidelines, and storm waves obliterate lines they cross, pushing material up the beach to a fresh line. The thickness and diversity of Alaska's Tidelines reflect the richness of ocean life, providing beach walkers a fascinating glimpse into the world beneath the waves.

An Alaskan Tideline is a colorful medley that can include red, green, and brown seaweeds torn from rocks or the seafloor, sticks and cones, grasses and needles, driftwood, pebbles and shells, fish bones, feathers, bits of bright red crab carapaces, the soft bodies of jellyfish, and even bones and carcasses of marine mammals. Many high-energy outer coast shorelines contain a jumble of giant logs, some escaped from logging in the Pacific Northwest. Where coal seams lie on the ocean floor, the line may include chunks of coal; near volcanic eruptions, the line includes pumice stones that float ashore. Following storms, Tidelines are often thickly piled, even with large tangles of bull kelp like rope tumbling it all together. Beached kelp fronds are often encrusted with colonial bryozoans, coralline algae, and invertebrate eggs, and driftwood may harbor gooseneck barnacles.

Tidelines may also include human-generated marine debris that washes ashore from all over the North Pacific; keeping Alaska's Tidelines clean of this debris is an ongoing effort. Lost or jettisoned fishing gear remains a problem, and other plastics, including single-use water bottles, continue to accumulate. Along with public awareness and conservation measures, regular marine debris cleanup efforts aid in reducing this threat to life along Alaska's shores. One continuing source of human-made material is the occasional loss of shipping containers from ships. As well, events in other parts of the Pacific bring debris to Alaska: debris from the 2011 tsunami in Japan continues to wash up along the shores of the Gulf of Alaska. The Tideline highlights how the entire North Pacific is part of one big interconnected ocean.

Tidelines are important for many other beings who scavenge them for food. Bears, river otters, wolves, martens, and foxes meander along Tidelines, picking through them for the remains of fish, crabs, and seaweed. Black-tailed deer and mountain goats visit beaches to feed on kelp, especially in springtime after a long winter without fresh greens. Gulls, bald eagles, sparrows, and other birds pick at the Tideline for food. Humans, too, scavenge tidelines as beachcombers, simply to see what washes up, or to collect shells or stones or usable pieces of wood.

Some beings live their whole lives in the Tideline. Beach hoppers, also called springtails, small shrimplike amphipods that can hop many times their body lengths, are important detritivores who eat bits of dead seaweed. With time, the material in the Tideline decomposes and adds nutrients back into the ocean. A Tideline is a world unto itself, full of life and many treasures, a world that changes with every tide.

LESLIE LEYLAND FIELDS

Tideline

Across the wide bay, fin whales feed,
great sinking ships.
Wind lifts ocean to lace.
Mountains wear their own sky,
Volcanoes fume.
The dizzying spruce sway shadows across the sun.
Under the bay, red corals grow houses
like veins, hearts.
And here, along the tideline, fragments of it all—
whale bones, ash, lost trees, homes.
Each time I come here with you
the continent's shelf tilts, empties, delivers
to our hands and feet this surplus.
And gathering these pieces

I am already generous,
            forgiving breached promises, lost homes, broken hopes.
I lay these weights down
on the beach,
now small and light
as the coil of red coral
I rest at your feet.

Sitka Periwinkle

Littorina sitkana

The Sitka Periwinkle is a small, round marine snail, about the size of a large pea or small marble. The shell, with four whorls, is highly variable in color, ranging from near-black to gray and brown with bands of white, green, orange, and yellow. The dark snail body has eyes, "antennae" called tentacles, and a mouth full of sharp teeth, and slides on a foot covered in mucus. Like other snails, periwinkles can close themselves inside their shells with "doors" called opercula—this prevents drying out when exposed at low tide. They can remain for long periods out of water, breathing air, and can suffocate if underwater for too much time.

The lovely name *periwinkle* is shared by a pretty purple flower (not native to Alaska), but the two names come from different roots. Our *periwinkle* derives from the Latin *pina* or Greek *pine*, meaning "mussel," and *wincel*, meaning "spiral shell."

Periwinkles live on rocks in intertidal zones and along rocky shores, on rockweed and other algae, in tidepools, and attached to pilings and docks, along the Alaska coastline as far north as the Chukchi Sea. They also live as far south as the coast of Oregon and as far across the Pacific as Siberia and Japan. Along low-energy shores, they can appear as dense as piles of pebbles. As grazers, they use their mouth parts—actually radulae, ribbonlike structures ("tongues") of tiny, horny teeth—to scrape diatoms and algae from rocks and other surfaces. The constant scraping is so abrasive that periwinkles wear away rock over time.

XTRATUF
XTRATUF
LSP

Eggs are laid in mucus mats, often by several females together. After hatching, the tiny snails settle very quickly and remain largely sedentary in their home places, usually moving less than a meter's distance in a month's time. This lack of mobility makes them vulnerable to environmental changes, heat, and desiccation. Pollution and silting are their primary threats. Sitka Periwinkle populations recovered very slowly after the 1989 *Exxon Valdez* oil spill in Southcentral Alaska—and even more slowly where the shores were treated with hot water washes to remove oil. Their ability to build and maintain strong shells may also be compromised by ocean acidification as the oceans absorb more carbon dioxide.

Periwinkles provide food for many other coastal beings, including fish, sea stars, other mollusks, birds, and especially crabs. Small hermit crabs sometimes use their empty shells for homes. Humans can also make a meal of periwinkles, after steaming or boiling a potful of them. Sitka Periwinkles appear on lists of foods eaten by Alaska's Indigenous Peoples, but in a land of many choice seafoods they aren't known as significant sources of nutrition. Instead, they're admired for their just-above-tide gatherings in wet crevices and along tidepools, where their multicolored whorls decorate the edges like lacework.

VIVIAN FAITH PRESCOTT

Dear Sitka Periwinkle,

I've always admired your four whorls, your spiral sculpturing with ridges and furrows, how your dark purple and gray bands catch light at low tide. I just wanted to say, I know you're used to spawning several times per year, but things have changed. There are predators you must be aware of—that sea star—don't be fooled, little globose one. There's a red rock crab behind every boulder and the bullhead is not your friend. Still, go out and enjoy those foggy days, eat and drink your fill of diatoms, algae, lichen and rockweed. Have fun in the splash zone! Oh, *Littorina sitkana*, may the eelgrass always keep you safe.

Northern Sea Otter

Enhydra lutris kenyoni

In quiet coves and sheltered bays, a rhythmic tapping ricochets off the shoreline: the sound of Sea Otters using rocks to crack the shells of sea urchins. Floating on their backs, their boat-shaped bellies a table, they feast, then roll repeatedly to clean their luscious dark brown fur.

An iconic and beloved sight in nearshore areas, playful Northern Sea Otters live along Alaska's shorelines from Southeast through Southcentral and throughout the Aleutians. Weighing in at 30 to 100 pounds, they're the heaviest member of the weasel family but among the smallest marine mammals.

The Sea Otter is the only marine mammal to stay warm without blubber, relying instead entirely on fur and the air trapped within it. Sea Otters have the densest fur of any animal on Earth; at a million hairs per square inch, their fur is fourteen hundred times more dense than human hair. This makes them well equipped for cold northern seas, but that same insulating fur keeps them from diving deep, so they're always in shallower waters, seldom more than three-quarters of a mile offshore.

They also survive in cold waters by consuming about a quarter of their body weight in food every day, generating metabolic heat. They eat more than a hundred different nearshore beings, including sea urchins, mollusks, crustaceans, and some fish. Using a rock to break the shells, they're among the few animals known to use tools. They store this rock in a loose pouch of skin under a foreleg. Otters also use this pouch to carry food to the surface. Rarely do they come ashore, not even to drink; their large kidneys allow them to process seawater.

Sea Otters are a classic example of a keystone species: both their presence and their absence affect their nearshore environment more dramatically than their size and numbers would indicate. Like Yellowstone's wolves, they've shown us how the return of keystone predators can restore habitat. Nowhere is this more obvious than with Sea Otters' impact on sea urchin populations and kelp: without Sea Otters, sea urchins can raze kelp forests to the ground. After Sea Otters returned to the coastline of Southcentral Alaska after being extirpated, kelp forests regrew and expanded. The longer that Sea Otters have held a keystone role in those ecosystems, the greater the expansion has been.

Not only do Sea Otters find food in kelp forests, but they often sleep among them, too. Rafting together by sex, using their front paws as eye masks during

their midday sleep, Sea Otters wrap themselves in kelp fronds or hold paws with one another to keep from drifting away.

Mothers tuck their pups in kelp while they dive for food. Pups, usually born in early summer, have a baby fur that makes them float like a cork, unable to dive. Not until they're several months old do they grow adult fur. So they spend those first few months atop their mothers' bellies, where the mothers constantly preen and fluff them, cradling them close to keep them warm and safe, and embracing them with a remarkable level of attentive affection.

Sea Otters figure prominently in Alaska's Indigenous cultures, particularly as reincarnated or shape-shifted family members. Frequently adorning totems, they're said to symbolize friendship, peace, family, and good fortune. In Tlingit stories, a Sea Otter saves a human from drowning and gifts him with a pouch of seeds from which all native trees sprout. In an Alutiiq origin story, Sea Otter was originally a man who was trapped by an incoming tide while collecting chitons. To save himself, he prayed to become a Sea Otter—a transformation that created all Sea Otters.

Sea Otters once lived from the Baja Peninsula northward in a long arc to the Russian Far East and Japan, numbering well over three hundred thousand, with Alaska as their central range. But intense hunting for their furs from 1741 to 1911 killed more than 99 percent of them, leaving barely a thousand worldwide. A 1911 hunting ban pulled them back from the brink of extinction and started their recovery. But because they can't dive deep, they won't cross major channels and don't disperse far on their own. So in the 1960s they were intentionally reintroduced to some areas. They now thrive in about two-thirds of their former range, and populations are stable or increasing except in the Aleutians, where their numbers have collapsed since the 1990s. The current theory for the Aleutian Sea Otter decline is that as whale, seal, and sea lion populations have fallen in those areas, transient orcas and sharks have had to switch to another food source—Sea Otters.

Alaska's Sea Otters were also hit hard by the 1989 *Exxon Valdez* oil spill, which killed thousands of them in Southcentral Alaska immediately; its lingering oil continues to harm their population. Oil spills present a significant threat to their existence, as the oil ruins their furs' insulating ability and then poisons them when they preen their oiled fur. Other threats include water pollution, shoreline development, ocean acidification, commercial shellfish harvests, and climate change.

The health of Alaska's shorelines—and the entire planet—depends upon Sea Otters in more ways than are easily visible. Sea Otters indirectly aid increased carbon sequestration through the health of those kelp forests their presence sustains, and which provide them habitat. And they delight us, as they raft together among the broad fronds, grooming themselves and the tiny fur-ball pups on their bellies.

SHAELENE GRACE MOLER

In Relationship

Mom always held me in the stern,
tightly. My hands were covered in
ocean spray. Waving with their feet,
otters greeted us on the way by.
Or at least that is what mom says.

They feast in kelp beds on shellfish,
cracking shells as skillfully as ravens.
Kelp houses otter, otter eats urchins,
kelp grows long and thick.
A forest.

Otters hold their pups
tightly, and give them to aunties
when their arms tire. They raft together,
moms and pups in one, males in another.
Raising youth in community, by community.

Beach Greens

Honckenya peploides

Beach Greens grow in brilliant yellow-green patches or mats in sandy or rocky areas high on most Alaska beaches (and, indeed, on beaches around most of the northern world). The plant is also known as Beach Chickweed, Sea Chickweed, Seabeach Sandwort, Seaside Sandplant, and Sea Purslane in English; Achaaqhluk or Atchaaqluk in Iñupiaq; and Itga'raleq in Yu'pik.

Welcomed as a vitamin-rich plant by generations of Alaska's Indigenous Peoples, Beach Greens were also sought by early western sailors and explorers to prevent scurvy. They remain a staple food for many coastal Alaskans today, for snacking on beaches, adding to salads and soups, and preserving for later use. The tender spring leaves eaten raw have a mild and lightly salty taste, and the cooked greens have been likened to cabbage. Later in summer, when they flower, their thicker leaves become tougher and bitter tasting.

Bears, deer, and foxes also like to eat fresh Beach Greens. These plants are especially important for wild beings in springtime, when they're in need of the vitamin-C-rich nutrition.

Beach Greens are distinct in the brightness of their waxy green leaves and in the shape and arrangement of their pointy opposing leaves on succulent stems. The greenish-white flowers are small and nearly hidden by the leaves. The scientific species name, *peploides*, is Greek for "cloak," referring to the way the leaves wrap around the flowers. (*Honckenya* refers to the eighteenth-century German botanist Gerhard Honkeny.)

Beach Greens are incredibly hardy and resilient. Their thick, succulent leaves and stems, along with the waxy coating that reflects light, help them not only retain water but also survive intense sunlight, strong winds, salt spray, and full immersion in salt water during higher tides. Low to the ground, with runners that take root at nodes along their stems, they can move over time to adjust to a beach's profile and tideline. Their long taproots keep them securely anchored in the sand. That anchoring, moreover, helps stabilize beaches, preventing erosion, and allows other plants to move in alongside them and eventually displace these cheerful green plants who shine in sunlight.

LAURELI IVANOFF

Making *Achaaqhluk*

We chopped. And chopped. And chopped.

We were chopping the skinny stalks of beach greens with my Gram's *ulus* at her table in Unalakleet, Alaska. Summer sunlight streamed through the window above the clean kitchen sink of her small HUD home, *The Price Is Right* playing on TV. The greens were a foot long and had leaves the size of a fingernail, which felt rubbery, like they'd squeak if I stroked them with just the right pressure. The stalks gave a good crunch when the *ulu* blade sliced through. We didn't say much.

Gram had picked the greens that morning from the beach just a short walk from her house, greens with an English name so literal—beach greens—I like to think they must have been named by a Native. My kids were with their dad, so I had all day to chop if I needed it. And it felt like I would: no matter how quickly I worked, the mountain of unchopped greens didn't seem to shrink. It felt like making a kale salad for a huge vegan wedding reception where only one dish would be served—kale salad. *Araa, this is going to take forever,* I remember thinking.

Now and then, Gram would take the bowl of chopped greens and grab a bundle with her hand to place in a pot of boiling water. As soon as the greens brightened, she transferred them to a one-gallon glass jar with metal tongs. When we finally cut the last of the greens, the container was full. Gram added some hot water until the greens were submerged, then topped the jar with a red-and-yellow cotton towel, tied in place with white cotton string.

The jar went into her dark backdoor entryway, where it sat for a month, undisturbed and fermenting. Once the greens gave off a hint of a sour smell, maybe a month later, they were done. She packed them into quart-sized Ziploc bags to freeze. Later, she would thaw a bag when she was ready to make *achaaqhluk,* a mixture of the fermented greens, blueberries, and sugar: the perfect dessert to serve after a heavy meal of dryfish, dried *ugruk* meat, potatoes, carrots, herring eggs, and seal oil.

Black Oystercatcher

Haematopus bachmani

A sharp, loud whistling erupts from the shoreline, and a pair of large soot-colored birds bursts skyward, alarm calls ringing louder than the waves. They settle onto a rocky outcropping, where their bodies are camouflaged against the dark rock. But their bright orange-red bills, long pink legs, and bull's-eye

yellow iris encircled by a red ring make Black Oystercatchers among the most conspicuous and exotically colored of Alaska's shorebirds.

They're also among the rarest shorebirds in North America. The total population of Black Oystercatchers along their range, from Alaska to Baja, is fewer than eleven thousand birds, of whom about 70 percent call Alaska home.

Ranging along Alaska's coastlines from the Aleutian Islands south, Black Oystercatchers spend their entire lives in view of the Pacific Ocean, on rocky shorelines and quiet bays. They are homebodies who mate for life, tightly rooted to place and to each other. Unlike most of Alaska's shorebirds, Black Oystercatchers don't migrate south in winter but gather in larger groups and find refuge from winter storms on mudflats close to rocky coastlines.

Oystercatcher allegiance to place gives them the distinction of being both a keystone species, since they're among the top predators in the intertidal zone, and an indicator species, as their presence indicates the relative health of the intertidal life on which they depend. Contrary to their common name, Black Oystercatchers seldom eat oysters—which don't grow wild in Alaskan waters. They are the ultimate tidepool enthusiasts, using their long flattened beaks to loosen limpets, periwinkles, chitons, mussels, marine worms, and other small intertidal life. They depend on tides; when waves wash over mussels and the bivalves open their shells, Oystercatchers take advantage.

An Oystercatcher nest is little more than a scraped depression high on a beach, adorned with a few shells, out in the open but nearly invisible. Nests are close enough to tidelines that they're vulnerable to sudden storms or wakes. Their eggs are hardy and can withstand repeated flooding; Black Oystercatchers will roll flooded eggs up the beach several times and still successfully hatch them. Parents noisily and zealously guard nests and chicks.

Year-round reliance on a narrow band of habitat for all aspects of life makes Oystercatchers especially vulnerable to human disturbances, whether increased kayaking and camping along their shores, boat wakes, coastal development, predation by introduced rats and foxes, marine debris, or oil spills. In Alaska, not only did the 1989 *Exxon Valdez* oil spill quickly kill 20 percent of them in the spill area, but the toxic oil also killed mussels and other intertidal life that feeds them, causing a sharp and continuing decline in their population. They're a species of high concern for both federal and state agencies, and for conservation organizations in the United States and Canada.

As distinctive and showy as they are, surprisingly little is known about Alaska's Black Oystercatchers. They still hold secrets, these year-round residents of Alaska's shores, like the pair with shoulders hunched, muttering to each other as they walk the beach in front of you. If they rise up with their piping call, creating aerial chaos, watch your step: they may be trying to lure you away from their well-camouflaged eggs or chicks hidden in plain sight just above the tide's reach.

KAYLENE JOHNSON-SULLIVAN

Unforeseen

With golden eyes circled by orange, a bright laterally
flattened beak, you look comically and perpetually
surprised, caught naked
with pink legs bare to the world.
Even your name is a ruse, since
you eat not oysters, but mollusk
and limpet.

An effervescent memory rises,
and I think of the summer the boys rowed a skiff

into the cove to catch fish. They befriended
a seal who twirled and pulled the rope
dangling from their boat. Eyes wide with
delight, they stomped up the cabin
steps for more herring to feed
their playmate. The seal waited
by the dock, then vanished
when adults came to look.

The boys have long since disappeared into
manhood, and I am the one surprised,
caught naked without them. They
were my truest home. For a while I
was bereft and bare, scoured by the raw
winds of a passing era.
What is there to do
except share this rocky shore with
creatures equally surprised?
What is there to do except laugh?

Aleutian Wild Rye

Elyleymus aleuticus

A member of the very large grass family, Aleutian Wild Rye is close kin to many other species of wild rye that grow tall and sturdy along coastal beaches in Alaska. Highly salt resistant, wild rye often grows where few other plants can and withstands being inundated with seawater at the highest tides. A single stalk may be three-quarters of an inch thick and consist of six or seven blades that unfurl upward to the height of an adult human. Along shorelines and hillsides they wave in the wind like ocean waves.

These long, wide, flat leaves serve as nesting material and the long, thick flower and seed clusters as food for many birds and small mammals. Beach ryes in general have long been utilized by humans for a wide variety of things, but they're especially useful for weaving into baskets, mats, socks, and other items. And they're extremely important to wave-washed and windy habitats as they provide erosion control on sandy and/or gravelly coastal areas and unstable dunes.

One thing that's special about Aleutian Wild Rye is its place in the artistry of Indigenous weavers, especially the Unangax̂ of Attu Island, who are known as makers of some of the finest grass baskets in the entire Native American world. Attuan baskets are renowned for their tiny, tight patterns made with strands so threadlike that the finished surface is sometimes compared to linen. The best grasses are typically harvested in summer just as the stalks are maturing and the heads of grain unfolding. They're found away from the beach, in meadows or on hillsides, where the growth of ferns around them forces them to gain height before the blades thicken. A lengthy preparation process follows, during which the select inner blades are split into threads with a fingernail.

The Attuan tradition of basketmaking was disrupted in 1942 when the Japanese attacked the Aleutian Islands and captured all forty-three Attuans, who were taken to Japan for the duration of the war. Nearly half died of starvation, malnutrition, and disease. At war's end, the survivors were not allowed by the American government to return to Attu. Their village, in any case, had been destroyed. No one lives on Attu Island today.

In 2017, a group of descendants of the Attuans returned to the village site. On the hillside, they collected armfuls of Aleutian Wild Rye to take back to weaver Agnes Thompson, who was born on another Aleutian island, Atka. Thompson belongs to a long line of weavers and learned the art from relatives born on Attu. Two of her baskets, finely woven from those collected grasses, now reside in the Museum of the Aleutians in Unalaska.

JERAH CHADWICK

From the Museum

In the old photograph
women shouldered their bundles
of grass gleaned from the hillsides
where it grows less coarse,
less exposed to salt winds
from the sea. Later they will

split and cure the grass
to trade for lumber, tools
to build the last church
on Attu. But for now they stand
a little stooped and smiling,
nameless as the makers

of these baskets lining
the museum shelf, Attu baskets
meant for fish, berries.
Some so tightly woven
you can't count the stitches
are fine as the cloth

those women embroidered
for the altar, in gold and greens
like the procession of hills
around their village

before their village was erased by bombs.

Daisy Brittle Star

Ophiopholis aculeata

This echinoderm's genus name, with *ophio* meaning "snake" and *pholis* meaning "spine," suggests its appearance—five long, thin, whiplike arms radiate from a central disk, and spines grow along the sides of the arms. The disk is less than an inch across, while the arms extend about five times that distance. These beings maneuver by flailing or rowing with their arms. The "brittle" in their common name acknowledges that an arm grabbed by a predator snaps off; the arm then grows back.

While closely related to sea stars, brittle stars are in their own class, represented by more than two thousand species. Although they're found around the world, mostly on the seafloor, the distribution of Daisy Brittle Stars is circumpolar. These beings live intertidally, often in rock crevices or among shells and seaweeds, and are the most common of the brittle stars.

In Alaska, the Daisy Brittle Star lives on the bottom of the Arctic and North Pacific Oceans but is usually spotted in tidepools or in the lower intertidal zone when a single leg extends from a crevice or from beneath a rock. The color varies but is often red and brown in a striped pattern. Other common names—Painted Brittle Star, Crevice Brittle Star, and Ubiquitous Brittle Star—may be more descriptive than "Daisy."

Brittle stars feel around for food—for the Daisy mostly microscopic organisms—and then use their tube feet to move bits along grooves in their arms to their mouths. Like all echinoderms, they have no head, no eyes, no brain, and no heart. They don't even have an intestine. Food goes from mouth to stomach, digestion occurs within stomach folds, and the bursae (cilia-filled sacs at the base of each arm) do the excreting. The bursae also contain gonads that launch sex cells into the water; a Daisy is either male or female, and the eggs and sperm meet up in the water for fertilization, then to become planktonic larvae.

Daisies provide food for other beings, including crabs, urchins, ducks, and sea stars. While they may be admired by humans, they're seldom collected for decorative purposes because of the brittleness that causes their arms to break off.

SUSANNA J. MISHLER

Brittle Stars

Nobody confessed; suddenly all of us,
even the innocent, wished we were the rocks we sat on

instead of feeling pressed onto them
like specimens smeared on a laboratory slide, so thin

light shone through our vessels
and revealed every deception—those given to us,

those we'd inflicted on ourselves and already, in fifth grade, we
squirmed. It was April, cloudy, and we wore coats

zippered to our noses in the damp wind.
Maybe that kid got bored and began by looking at one,

fingering its gnarled rays and wondering if they were actually
brittle.
All day we felt the sand grains stuck between our toes from
running

in the Skeleton Forest. The trees, we were told, sank in an
earthquake, and we all knew without asking

which one. "It was so big,"
our parents said, "it split streets open, sank the coastline."

We emptied our socks. The skeletal parts of the world,
at least, seemed safe from us.

For each one to grow to the size of a human hand
takes years. I don't know

if that kid stacked all fifteen on top of one another,
their rays tangled like a green crayon drawing of the sun,

and stood with a rock as big as he could lift
or if he knelt by them, turned each to peer in its stomach,

then placed them on his anvil one by one by one.
We walked the beach, listening. We could hear

a chorus of small things calling.
At sunset we prowled the waterline for octopus dens,

found blood stars and leather stars instead,
clinging to the dark sides of boulders.

When the *Exxon Valdez* hit Bligh Reef,
the captain was drunk and the oil
spread its sticky cape.

                                        Captain,
there are stars out there bigger than our hands.
Any child can spend an evening

with his back in sand and hands held to the sky,
eclipsing the lights of planets with his thumbs.

Ghost Forest

Alaska's landscape is incredibly dynamic. Mountains push up and slide away, glaciers surge and retreat, coastlines erode, volcanoes swell and build new lava domes. The land itself, the crust over core, lifts and subsides. In much of the state, the land is still rising after Ice Age icefields melted, ending the pressure of their great weights. That isostatic rebound (also called postglacial rebound) in and around Glacier Bay is occurring at the fastest rate on Earth, with an uplift of more than 18 feet since the glacial retreat that began about 250 years ago. Elsewhere earthquakes lift, lower, and tilt the landscape.

During the Great Alaska Earthquake of 1964—the largest recorded quake ever to hit North America—vast stretches of the coast near the epicenter in Prince William Sound tilted, lifted, or subsided. In the northern sound, the land sank as much as 6 feet; in the southern sound, land rose up to 38 feet, destroying clam beds as far east as the Copper River Delta.

Boat through parts of the sound today and you'll discover stands of dead, skeletal trees along the shore. Their barren and broken grayness contrasts

sharply with the deep forest-green behind them. Eagles and gulls keep watch from their gaunt upper limbs. You'll see the same along Turnagain Arm south of Anchorage, where the community of Portage sank 8 feet, became a saltwater marsh, and had to be abandoned. In Kachemak Bay, a low-lying peninsula features the same gray-stick landscape.

These are Ghost Forests. When bedrock dropped, all the clay, soil, mud, and vegetation on top fell with it. Salt water inundated the ground, was taken up by tree roots, and poisoned the trees. Salt is a great preservative, drawing moisture from the wood and making the trees inhospitable to fungi and insects that cause rot. Even sixty years after the great quake, Ghost Forests are still stiffly standing, still ghosting over the land. The roots and lower sections of trunk have begun to trap mud and organic material to gradually build up new layers to support spreading plant life.

Other, older Ghost Forests, some buried in mud or earth or reduced to stumps, help scientists determine the dates and extents of earlier seismic events—frequent (in geological time) in Alaska and other parts of the Ring of Fire where tectonic plates collide. Another category of Ghost Forest are forests that were overrun by glaciers; such forests, buried in gravel ahead of advancing glaciers and later sheared off by advancing ice, were eventually uncovered by glacial retreat and erosion by rivers.

All over the world, Ghost Forests are now increasing in number as the sea level rises and seawater invades coastal areas. These provide a visual indicator of change related to our warming world.

RAY BALL

Field Guide to the Ghost Forest

on a wisp

,

of a branch
/

that will not /
bud, a brambling rests.

migratory
,
it does
not remember
the calamity of salt.

ground abandoned
its purpose. sea rose
to archive forest

.

spruce roots

/

submerged into
quaking silt.
petrified remains
                of briny bark
                        .

birdsong calling
a braided shroud

&

the living still
among the dead

.

Steller Sea Lion

Eumetopias jubatus

In white-capped seas or quiet bays, at any moment the waters may erupt with the bristled snout and powerful porpoising of a Steller Sea Lion. One tosses a silver salmon high into the air, then catches it in a wide red mouth. Another bursts up onto a rocky islet, using powerful front flippers to clamber up high.

Enlivening Alaska's coastal waters as far west as the Aleutians and as far north as the Bering Sea, Steller Sea Lions live in an arc around the North Pacific, from central California to Japan. They are named after Georg Wilhelm Steller, a naturalist on the 1741 Vitus Bering expedition. (Steller's name is also attached to other Alaskan beings, like the Steller's jay and the Steller's eider, a remnant of Russian exploration and occupation.) Alaska's Indigenous Peoples have their own names for this being, some of which translate to "sea wolf." In Alutiiq, the Steller Sea Lion is known as Wiinaq; in Central Yup'ik, Uginaq (sometimes Apakcuk); and in Siberian Yupik, Ulgaq.

Steller called them sea lions because the males have muscular chests and shoulders covered by long, coarse hair that looks like a lion's mane, and for their sound. Stellers, like African lions, roar. Not for them the bark of the California sea lion; Stellers, especially at their rookeries and haul-outs, emit a chorus of low-pitched belches, growls, and snorts mixed with the higher bleats of pups. Stellers also make underwater clicks, barks, and belches.

Adult male Steller Sea Lions are huge: beachmasters (dominant individuals) can weigh up to 2,500 pounds. Females, meanwhile, reach only 800 pounds. Like walrus, Stellers have long whiskers for sensing prey and feeling their way underwater. Their coats are tan to reddish brown, except for pups, who remain dark brown until their first molt.

Stellers are the largest members of the eared seal family, called otariids. This family includes California sea lions, who range into Southeast Alaska, and northern fur seals, who, hunted to near extinction by Russian fur hunters, have their last stronghold on Alaska's Pribilof Islands, where two-thirds of their remaining world population breed and pup. These eared seals differ from earless seals, phocids, by having visible external ear flaps and long hind flippers that can be turned under and allow them to move about easily on land.

In Alaska, Steller Sea Lions inhabit the same areas as harbor seals, but they're easily distinguished by size, appearance, and behavior. The smaller harbor seals, with large round eyes, quietly pop up out of the water, while Stellers jet to the surface with audible snorts, fearless and curious. And like all earless seals, harbor seals can't climb high; instead, they slip up onto rocks as the tide recedes, and pup on icebergs. Steller Sea Lions are capable climbers; they forcefully clamber onto islets and rocky headlands, and drape across angular rocks in groups, their thick blubber cushioning any surface.

Steller Sea Lions need land and sea in equal measure. Highly social, they rest together at haul-outs, bodies overlapping, and they mate, give birth, and nurse their pups at rookeries to which they return every year. A few days after giving birth, females must take to the water to feed; when they return, they find their pups in the crowded, noisy rookery by distinctive scent and sounds—each pup and mother have a unique call for each other. These rookeries are fiercely protected by the Sea Lions and are particularly sensitive to disturbances, which may cause stampeding and kill females and pups. Federal law prescribes safe distances to prevent harassment by boats and humans.

At sea, Steller Sea Lions often travel in small groups, gathering in rafts near good feeding areas. Sometimes a Steller Sea Lion will hold one flipper out of the water as if waving. The flipper, a less-insulated body part, absorbs heat from the air, which is then circulated to the rest of the body. Their curiosity is well known: king crab fishermen put a solid-core "sea lion buoy" on their buoy lines, as Stellers often bite and pop inflated buoys.

Steller Sea Lions eat more than a hundred kinds of fish, as well as squid and octopus; they largely forage along the shallow continental shelf, although they are able to dive to more than 1,000 feet. Females, and thus pups, are particularly sensitive to food supply, as females have high nutritional needs; while nursing, they can't stray far from rookeries to feed, and they're often eating for three: nursing one pup while pregnant with the next.

Historically, Steller Sea Lions were abundant throughout many parts of the coastal North Pacific Ocean, numbering well over three hundred thousand in the 1960s. Indigenous Peoples hunted them for their meat, hides, oil, and whiskers; their images appear frequently in coastal Indigenous art. However, their western population decreased by more than 80 percent from the 1970s to the early 2000s. Both the western and eastern populations were listed as endangered in 1990. The eastern population was delisted in 2013, and the western population began to recover in the early 2000s in some parts of the Aleutian Islands. A lack of high-quality food is considered the primary cause of the decline continuing in other parts of their range.

Like other populations of marine mammals, Stellers are at risk from a shortage of the more nutritious fatty fish and larger sizes of fish, some of which are also targeted by large commercial fishing operations. Populations of fish are changing as well because of warming and acidifying seas, and Sea Lions suffer from the accumulation of toxic substances in their bodies. Steller Sea Lions also die from entanglement in fishing gear and illegal shooting.

Curious, fearless, social, and highly intelligent, Steller Sea Lions knit land and sea. They indicate ecosystem health and—everywhere along Alaska's shores—animate their world with their exuberance.

BONNIE DEMERJIAN

See, Lions

It's just another day.
The usual lowering clouds obscure the sun again,
a breath of wind to frill the water echoes sulky sky.

At the window I consider chores to choose,
mid-morning coffee near at hand, when wonder elbows in—
three blunt snouts cleave the rimpled waves.

Three bearers of mystery pierce the skin of ordinary.
Three prophets armed with fiercest teeth and brawny power.
They've come to grip the day, to thrust astonishment into our
dazzled eyes.

Glacier

Tidewater, hanging, cirque, piedmont, icefield: all of these types of Glaciers are found in Alaska, home to about 27,000 of them, 664 of whom are named. Every one is unique and constantly changing. They surge, they retreat, they slide over and crumble rock, shoving rock debris to form lateral and terminal moraines. They jackknife into salt water with resplendent songs. They groan and crack and roar.

Glaciers are ecosystems unto themselves, home to ice worms, pink algae, glacial midges, snow fleas, rotifers, and glacial copepods, upon which birds like snow buntings come to feed. Even their cryoconite meltwater holes contain complex ecosystems of microbes like bacteria, algae, and viruses.

Where Glaciers pour into salt water, they cause a nutrient upwelling that feeds kittiwakes, humpback whales, harbor seals, and Kittlitz's murrelets, who are almost entirely dependent on tidewater Glaciers. Their icebergs and ice floes shelter life, too: harbor seals and sea otters give birth and rest on them.

The layers of a Glacier contain the history of Earth's atmosphere, giving evidence of volcanic eruptions and windstorms as well as human activities such as industrial pollution and nuclear weapons detonation. They are like frozen time capsules. In Alaska, cores drilled from Glaciers of the Alaska Range document five thousand years of past temperatures, precipitation rates, and atmospheric patterns—which can help scientists estimate future effects of climate change in the region.

According to the Tlingit, Glaciers are either male or female: male Glaciers advance, while females retreat, and each is proffered different rituals and respect. In recent decades, many male Glaciers have become female, retreating farther from us each year. In fact, more than 95 percent of Alaska's Glaciers are now stagnating, thinning, and retreating.

Malaspina Glacier, covering 1,500 square miles at the top of the Alaska Panhandle, is the world's largest piedmont Glacier (a Glacier that spills out of mountains onto flat land and forms a lobe) but is now thinning and pockmarked with meltwater ponds. Prince William Sound's Columbia Glacier, one of the most studied Glaciers in the world, was stable until 1980; since then, its terminus has receded 12 miles and no longer reaches the sea.

Melting Glaciers determine freshwater abundance and the salinity of surrounding seawater, which affect the beings dependent on what have been more stable conditions. As they depart, Glaciers also leave behind U-shaped valleys,

fjords, islands, and hanging valleys—new (or returning, in geological time) habitats that provide homes for different or expanding populations of beings.

Over an eighteen-month period beginning in late 2003, South Sawyer Glacier in Tracy Arm-Fords Terror Wilderness retreated 1.5 miles, its 300-foot face slumping and receding. Within a year, the new curve of bay filled with ice floes and harbor seals, and a newly revealed rock mound nearly levitated with a thriving colony of arctic terns.

JEREMY PATAKY

Glacier Requiem

Glacier dressed in late 20th century finery
you were not snow white
but a bit ashen, a tad dirty.

Dressed like a one-hit-wonder singing
Baby don't go tunes.

I can hum the melody of your
single but will never remember the lyrics.

I drank you,
some from my hand, some from my flask
that also works with whiskey
from the city
distilled with water from the lake that catches
all that Eklutna Glacier melts

catches rain and undone snow as well, no doubt.
Catches stray reflections of golden eagles migrating

and hapless wildice skaters sailing solo on its windswept ice.

One flies clean into an open water trap
like a mosquito into a biker's open maw—

there's no way back from those black holes.
And there's no way to get you back, glacier.

Those summers we'd step onto you from land up *there*.

That year when we could see, for once, the ridge beyond from *here*.

The new map with new lakes grown before you
like entire valleys freed up in the warm spells,

fattening the creeks and rivers
sloughs and breaks
broaching the beachfront questions
questing from alpine vantages down

to shorelines and attentive, pushy oceans.

The confident seas
seizing all the capes
and surging up
spring-tide-sloppy streets
or fondling ships that chaff at dock lines.

You deserved a better afterlife
but you'll be waters staining carpets in the bedrooms of the poor,
the poor plus indignant monied ones
still soaking in the notion that is no notion but is proof, in fact,
is proof

that the shoreline as we thought we knew it
shrunk—

Glacier, I mistook you for a glacier
primped for parties yet uncalendared—

Party of one, I am, without you, now that you departed.
I'm planted by your grave
of gravel in a valley growing brush,

breath stopped by a throat-lump plug
of ice that I would melt,
if I could, myself, in time, sometime,

if the world that we made wasn't set and eager

to do it for me. To do it to me.

Marbled Murrelet

Brachyramphus marmoratus

Kittlitz's Murrelet

Brachyramphus brevirostris

Bobbing on the sea in pairs, they may at first look like tiny brown buoys. Then, with a sudden flick of their stout wings, they disappear underwater. Within seconds, they pop back up to the surface. Murrelets—both Marbled and Kittlitz's—spend summers in nearshore waters, where they dive after zooplankton and small fish.

These small, squat seabirds fish quiet waters in Alaska, the Marbled in Southeast and Southcentral Alaska and the less common Kittlitz's in a more limited range, in parts of Alaska and the Russian Far East. In winter both typically migrate to waters farther offshore. The two can be told apart only with a discerning eye, especially as their plumages change with the seasons. The Marbled has a marbled appearance in summer and a starker black-and-white coloration in winter. The Kittlitz's (named for the German scientist who first collected one on a Russian expedition in the 1800s) has a shorter bill and shows white in the tail when flying and above the eye in winter plumage.

Marbled Murrelets are birds of two worlds: they're tree-nesting seabirds who feed on small forage fish like herring and sand lance, and then travel up to 50 miles inland to nest in old-growth forests. They are, in fact, the only

member of the alcid family (which includes murres, guillemots, puffins, and auklets) who nest in the coastal temperate rainforest. In particular, Marbleds nest high up on the broad, mossy limbs of mature trees two hundred or more years old, where the female lays a single egg. Although they were first described in the 1700s, it was only in 1974, some years after loggers in British Columbia reported finding two chicks on the ground after felling a western red cedar, that a tree surgeon trimming storm-damaged branches in Big Basin Redwoods State Park in California discovered a fuzzy Marbled Murrelet chick in a nest high in an ancient Douglas-fir. Flying straight from sea to tree, Marbleds may go their entire lives without their webbed feet touching land. Their flight is bullet-fast; they've been clocked at over 100 miles per hour.

In the Pacific Northwest, these birds are listed as threatened under the Endangered Species Act due to habitat loss from logging. Although they're still common in Alaska, which is home now to about 75 percent of their remaining world population, they're thought to be in decline here due to climate change and shifting forage fish dynamics. Because of their dependence on both terrestrial and marine ecosystems, Marbleds are important indicators of the overall environmental health of coastal ecosystems.

The Kittlitz's Murrelet is even more mysterious than the Marbled, and knowledge of their ecology remains limited. These Murrelets were only recently found by ornithologists to be nesting on treeless, rocky mountain slopes and rockslides, shaped by glacial action (although Indigenous Peoples had already known this). Like the Marbleds in their trees, they raise one chick. These beings feed in cold, turbid waters, especially in front of tidewater glaciers. As fresh glacial meltwater wells up from the depths, bringing nutrient-rich waters to the surface, plankton blooms, small fish thrive, and Kittlitz's Murrelets feast alongside kittiwakes and harbor seals. Because of Kittlitz's Murrelets' dependence on tidewater glaciers, they've been nicknamed "glacier murrelets."

Little is known of population numbers, although Kittlitz's are listed by the International Union for Conservation of Nature as near threatened and as a candidate for listing under the Endangered Species Act. Major population centers in Alaska's Prince William Sound, Glacier Bay, and Kenai Fjords have experienced declines in recent years, likely due to receding glaciers and altered water flows as these glaciers no longer calve into the sea. The Kittlitz's Murrelet has become a poster bird for the damage caused by global warming. This bird is also vulnerable to oil spills; an estimated 10 percent of the world population died in Alaska's 1989 *Exxon Valdez* oil spill.

These small, unassuming-looking cousins, whose secrets we're slowly learning, knit together land and sea. The Marbled needs the slow-growing moss platforms of ancient temperate rainforest trees, and the Kittlitz's needs rocky slopes near tidewater glaciers. And both require healthy, thriving seas.

SHAUNA POTOCKY

Mystery of the Murrelets

Your secrets—
some quiet pact
made millennia ago

with the crags, the talus,
with the deep forest, old trees

who, content, hide
your precious mottled young
while determined birders

hardly find you
in the canopy of spruce
or among rough rock.

You sojourn from sea
to nest—to lay, to fledge
your young.

You, full of story lines
we cannot translate.
Your hints miss us.

So we stay astounded,
wild-eyed at your sighting,

hoping to see
where you came from—
where, next, you will go.

Gumboot Chiton

Cryptochiton stelleri

Also known as the Giant Western Fiery Chiton or Giant Pacific Chiton, the Gumboot Chiton is the largest chiton in the world, growing in Alaska to lengths of up to a foot. Eight armored, butterfly-shaped plates cover this marine mollusk's flexible back; unlike those of other chitons, the plates are hidden by a red leathery skin.

More than forty chiton species live in Alaska waters and nearly a thousand in the world. Alaska's best known (and most utilized) are the Gumboot and the much smaller black Katy chiton or bidarki (named after the Russian term for a small boat, which this being resembles).

The Gumboot's common name comes from the being's resemblance to a rubber boot sole. The Latin name means "Steller's hidden chiton," recognizing both the hidden plates and the species' first scientific description by Georg Wilhelm Steller, the German zoologist who accompanied Vitus Bering to Alaska in 1741. Among Alaska's Indigenous Peoples, the Tlingit know Gumboot as Shaaw and the Tsimshian as 'Yaanst.

The Gumboot's underside is orange and consists mostly of the foot (similar to a snail or slug foot), with gills at the edges. The mouth, on one end, is equipped with arrays of fine, very sharp teeth on a tonguelike radula for scraping algae from rocks. Gumboots eat red and green algae, as well as some kelp, and move slowly through intertidal and subtidal areas. Adults usually do not travel very far—perhaps 3 feet in a day and within a range of 60 feet during a year.

In early summer, male and female Gumboots release eggs and sperm into the water. The fertilized eggs hatch into larvae who swim for just a day before settling onto the rocks where they will live out their lives, which can be as long as twenty-five years. Predators include sea stars, crabs, fish, and sea otters—as well as humans. Sea stars lift chitons from rocks, then use their tube feet to keep the chitons from protectively curling up into their shells. Gumboots are also vulnerable to storms, pollution, and overharvesting. Because they're long-lived, they don't easily recover from population declines.

In Alaska, Gumboots live from Southeast through the Gulf of Alaska to the Aleutians. Tlingit, Haida, Tsimshian, Alutiiq, and Unangax̂ all eat chitons as traditional foods. Although Gumboots are tougher than black Katy chitons, and the bad-tasting skin needs to be removed, they're often favored for their large size; they're also easy to find on rocky beaches during low tides at all times of

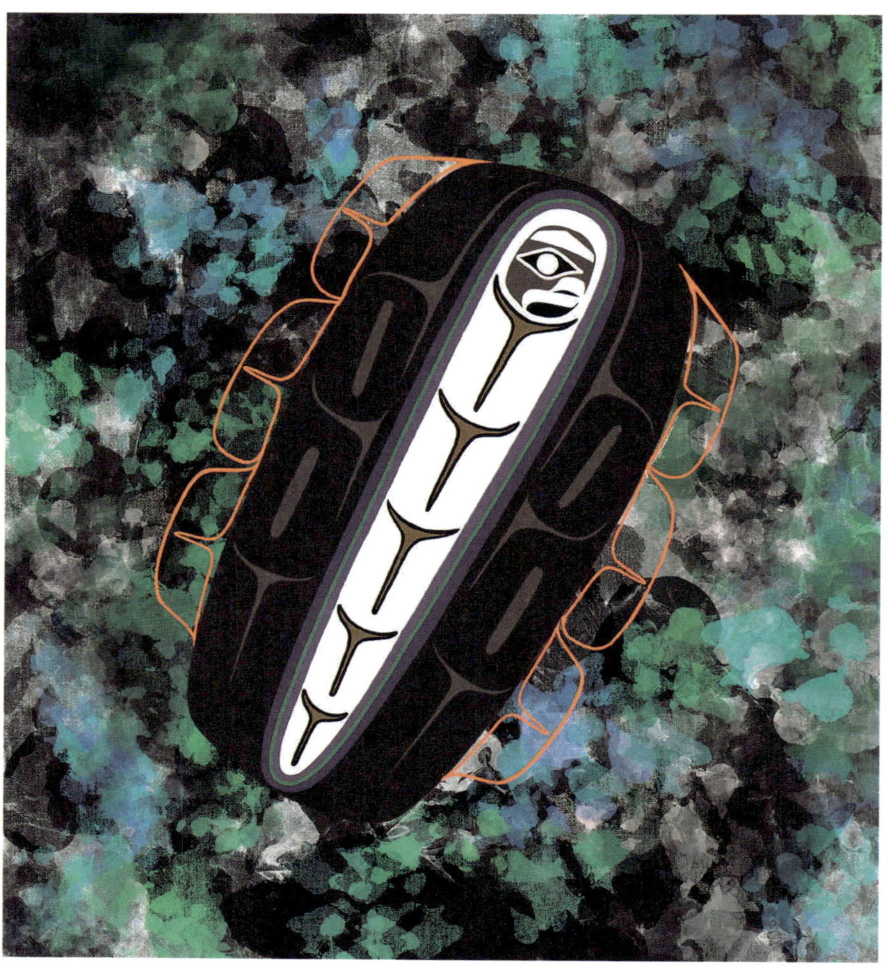

year. Harvesters pry them from rocks with knives or other sharp instruments. Chiton shells, whole or separated into plates, can be saved for use in decorative jewelry, especially earrings.

These nocturnal grazers are important members of their intertidal community: not only are they food for many beings, but they also help other intertidal life by keeping algae in check. They're also a wonder of the natural world for their ability to live both beneath and above the sea. When submerged, they breathe, like fish, through their gills; when exposed to air during low tide, they, like us humans, breathe oxygen.

ISHMAEL KHAAGWÁASK' HOPE

Shaaw

Flicked off with a butter knife.
Thanked
for giving itself.
Told it will feed us.
Pickled and jarred.
Nutrients of the fire stories.
The lowtide rocks.
Drawn to life from the tidewater.
The unlaying dawn.

Eelgrass

Zostera marina

Beneath glassy waters in protected bays, inlets, and lagoons, shimmering green ribbons bend and sway with the currents, hiding a diversity of marine life. When the tides recede, these ribbons lie flat, providing wet refuge for an array of invertebrates.

Eelgrass, the world's most widely distributed marine flowering plant, thrives in the soft sediments of shallow coastal waters, forming a vital part of the greater ocean ecosystem. In their dense aquatic meadows, Eelgrass provides safe harbor and forage to a huge diversity of marine beings, including octopuses, sea stars, anemones, plankton, crabs, and an array of fish.

Juvenile salmon and cod feed and hide in their meadows; the eel-like Pacific sand lance uses a pointed snout to burrow in the soft sediment Eelgrass holds in place; Pacific herring eggs attach to the long, thin blades. Seals, otters, and sea lions forage and hide in these waving underwater forests. Waterfowl consume the green blades; Eelgrass is the primary food for Pacific black brant, who feast on them both along Alaska's southwestern coast in spring, summer, and fall, and in wintering grounds in Baja California.

Eelgrass meadows are among the world's most widespread and productive ecosystems, vital not only for biodiversity but also for carbon sequestration. Eelgrass and other coastal "blue carbon" ecosystems capture and store CO_2 and methane up to thirty-five times faster than tropical rainforests. Eelgrass meadows also buffer coasts from storm erosion, moderate the effects of ocean acidification, and improve water quality by absorbing and trapping pollutants, including PCBs, algal blooms, and sediment.

Worldwide, Eelgrass meadows are declining due to dredging, pollution, coastal development, and sea-level rise, prompting efforts to protect and replant these meadows in locations such as Puget Sound, the Chesapeake Bay, Denmark, and Scotland. For now, Alaska's Eelgrass meadows are thriving and vast, making them even more important globally. Izembek Lagoon, near Bristol Bay, contains the largest Eelgrass meadows in the world—more than 65 square miles of undulating green fronds harboring a tapestry of life.

CAROL HULT

Ode to eelgrass

You sway in the currents
of this quiet cove
as I paddle, slowly, over
the dense forest of your long green blades
waving like ribbons from the deep. The tips,
just below my kayak, ripple
in the clear, cold water,
reaching for the sun.

Zostera marina, though you grow in the salty sea
you are not seaweed. Grass of the ocean,
with roots in soft seabeds of mud, sand, gravel, you live
in nearshore waters and give refuge to countless creatures.
Snails and hermit crabs hide in your beds. Herring eggs cling
to your blades. Young salmon swim in your meadows.
Wintering geese feed on your leaves.

On a full moon's low tide you lie exposed on the beach
among sea stars, bull kelp, barnacles, a tangle of life all
waiting for the sea to return and when it does
you lift upright in praise of water and light.

Oh, humble eelgrass, you are beautiful
swaying peacefully in the ocean.
Yet you are so much more,
a blue carbon ecosystem in
a world that needs to let you be.

This watery earth is your home.
You favor no one, you harbor all.

Temperate Rainforest & Coastal Mountains

Stretching along a thousand-mile arc of the Gulf of Alaska shoreline, the coastal forests of Southeast and Southcentral Alaska are composed predominantly of Sitka spruce and western hemlock, with pockets of yellow cedar, and are classified as temperate rainforests. Like tropical rainforests, these forests are characterized by a closed, multilayered canopy; moisture-dependent vegetation; the presence of epiphytic beings like algae, mosses, lichens, and ferns; and the absence of wildfire.

Temperate rainforests are cooler than their tropical counterparts but receive about the same amount of precipitation—up to 200 inches of rain per year. The density of living plant life—up to 1,000 tons per acre—is the greatest of any forest in the world. The trees, other plant life, and the soil play an essential role in sequestering carbon and thus mitigating climate change.

Globally, temperate rainforests are rare: they cover only about 2.5 percent of the area of tropical rainforests and only about 0.1 percent of Earth's land surface. Virtually all temperate rainforest is found on the Pacific coasts—along the northwest coast of North America and the coasts of southern Chile, New Zealand, southern Australia, and Japan. Alaska contains the northernmost reach of the temperate rainforest that extends south to northern California.

Located between the stormy Gulf of Alaska and the highest coastal mountains in the world, Alaska's coastal rainforest forms a thin living membrane between sea and ice. The glaciated coastal mountains divide into three ranges: Coast, St. Elias, and Chugach. Many of these ice-laden peaks, with jagged spires and ridgelines knifing above icefields and glaciers, rise steeply. Mount St. Elias, in Wrangell–St. Elias National Park and Preserve, is the second highest mountain in North America at 18,008 feet. It stands on the Alaska-Yukon border just

10 miles from the ocean at Icy Bay. The icefields ringing these mountains are the most extensive nonpolar icefields in the world.

Water is the lifeblood of the coastal rainforest ecosystem. Alaska's coastal mountains shape the climate that sustains the rainforests by forcing moisture-laden air to rise and then precipitate onto their western slopes and forests. As humid air from the gulf climbs these steep slopes, it releases hundreds of millions of tons of water each year. This water is filtered by the forest and drains down thousands of creeks and rivers as it flows back to the sea. Rainfall in Alaska's coastal forest is so heavy that an extensive lens of low-salinity water flows atop saline ocean waters west along the Alaska Coastal Current toward Kodiak and the Aleutian Islands, then northward along the west and north coasts of Alaska—in essence, a freshwater river on top of the salty sea.

The forest maintains stream water temperature, nutrient balance, and flow rate. Organic terpenes dripping from coastal conifer needles enhance the phytoplankton productivity of coastal ocean waters. Growing upon deglaciated, poorly developed soils over bedrock, tree roots concentrate on the shallow forest floor, predisposing them to windthrow during regular hurricane-force storms. In old-growth forests, this adds rotting logs to enrich the forest floor, and the forest becomes multistoried with a wide variety of age classes and many live, dead, and dying trees.

Most of Alaska's rainforest today lies within the boundaries of the Tongass and Chugach National Forests. Some old-growth trees here can be 12 feet in diameter, over 200 feet tall, and more than a thousand years old. Interspersed with wetlands, peat bogs, meadows, and alpine areas, the coastal rainforest forms an elegant patchwork of habitats for at least four hundred fish and wildlife species.

These forests shelter and cool salmon streams and provide woody debris that creates pools in which fish rest and spawn, supporting all five species of Alaska's salmon. Millions of salmon returning each year in turn enrich the streams with the nutrients they've brought from the ocean and feed many other beings, from invertebrates to brown bears to plant life and trees. Some research has found that trees on the banks of salmon rivers grow faster than trees on the banks of rivers without salmon.

The northern flying squirrel, who relies on specialized flaps of skin to glide for up to 100 yards through the forest's canopy, is just one of at least forty species of land mammals who live in the forests. This squirrel feeds primarily on the richness of the forest's lichens and fungi. Some beings move between coastal mountains and forest; mountain goats descend to coastal forests to escape heavy winter snows higher in the mountains.

Less obvious interrelationships occur on the microscopic level, among algae, yeasts, and bacteria that live on tree needles and support tiny grazing invertebrates that then feed others. Within the forest floor, tree roots and mycorrhizal fungi also enjoy a vital relationship in which fungi help trees to absorb water and nutrients, and trees nourish fungi.

And then there are the birds! Dozens of species—like warblers, brown creepers, goshawks, and grouse—rely on these forests and the mountain meadows that rise from them.

Historically, the Tongass National Forest and Alaska Native corporation lands have been managed mostly for logging. It takes more than 250 years for a logged temperate rainforest to regrow with the multi-aged trees and understory characteristic of an ancient forest, although even then the complexity is not replicated.

Alaska's coastal temperate rainforest and mountain ranges hold some of the last remaining old-growth stands of this globally rare and rich habitat. Above all, there's the sheer beauty of it: the stunning movement from deep blue seas to dense green forests to glittering icefields and soaring, rocky peaks.

Rufous Hummingbird

Selasphorus rufus

The feisty little Rufous Hummingbird is the most widely distributed hummingbird in North America and the only one commonly found in Southeast and Southcentral Alaska. (Anna's and Costa's Hummingbirds are infrequent visitors.) This being has the northernmost breeding range of any hummingbird and makes the longest migration of any North American hummingbird, flying all the way to the southern US and Central America for the winter. The Indigenous Peoples of Southeast Alaska know this hummingbird, Dagatgiyáa in Tlingit, as a messenger of joy, honored on family crests and in both traditional and contemporary art.

Hummingbirds are biological wonders, and not just for their shockingly glowing colors—a color diversity exceeding that of all other bird species combined. These smallest of warm-blooded animals, who live only in the Americas, also have the fastest wingbeats and heartbeats of any bird; what's more, they're the only birds who can fly straight up and down and backward. It's no surprise, given all this agility and speed, that they often consume half their body weight daily in food.

The Rufous Hummingbird shares all these traits, including the intensifying, shifting colors. The male Rufous has bright orange on his belly and back and an iridescent red throat (known as a gorget) that he flashes to either warn or impress other Rufous Hummingbirds. The female's colors are more muted, with green above, red-brown (rufous) flanks, and often a spot of orange on the throat. They are both about 3 inches tall and weigh less than a nickel.

Despite being so small, Rufous Hummingbirds are tough little birds. They survive cold weather, even freezing temperatures and blizzards, by going into torpor (a sleeplike state) to save energy. In fact, they routinely go into torpor during cool or cold Alaskan nights and have been seen in torpor at feeders, hanging upside down.

By darting from one flower to another to sip nectar with their long, slender bills and very long tongues, hummingbirds act as important pollinators. Rufouses' favored flowers are colorful and tubular and include Alaska salmonberries, yellow paintbrush, fireweed, and currants. They may visit up to five thousand blossoms in a day. They also consume protein and fat from insects, including gnats and midges caught in the air and aphids taken from plants.

Rufous Hummingbird females build their nests on drooping branches up to 30 feet high in a variety of coniferous or deciduous trees. In its finished form, the nest is a miniaturized wonder about 2 inches from edge to edge, with its inner cup about 1 inch across. It's built of soft plant material knitted together with spiderwebs, and the outside is camouflaged with bits of lichen and moss. The female lays two or three tiny white eggs. She feeds the nestlings by sticking her bill and tongue into their mouths and regurgitating insects, perhaps mixed with nectar.

In flight, the wings of a male Rufous emit a metallic whining sound, but it's the aerial courtship display that is most dramatic. The male makes a series of steep, J-shaped dives that begin around points on a circle, very fast in descent (up to 60 or 70 miles per hour) and slow on the climb, all the while stuttering a *dit-dit-dit-deer*. The wingbeats of the Rufous are incredibly fast—between fifty-two and sixty-two each second. During migration, these beings fly at 20 to 30 miles per hour and can cover 500 miles at a time.

Although the Rufous population overall is in decline (down 67 percent between 1966 and 2019), their numbers in Alaska are growing, likely due to the warming climate. Most hummingbird populations are in trouble, as are many pollinators, because climate change is shifting the flowering times of the plants they depend on. Banding programs have found that Rufouses return to the same areas each year, even to the same flowering plants and feeders. These banding studies have also determined that only about 60 percent of Rufous Hummingbirds survive their annual migrations. Threats to them include climate change, invasive species, habitat loss, and pesticides.

Perhaps the most magical thing about hummingbirds, including Rufouses, is the way their brilliant colors intensify and transform depending on the angle of view. This effect comes from tiny structures in their feathers—different from the feathers of other birds—that capture, bend, and reflect sunlight. The male Rufous's throat can change from black to red to orange, yellow, or lime green. It's a delight to see this spark of light in all his changing colors—whether at a feeder, in a garden, or among wildflowers—as he darts and hovers and zooms away.

DON REARDEN

Aurora in the Midnight Sun

As if they escaped from a winter
night sky, these traces of joy zip
from one salmonberry flower to the next.
Colorful confetti in perpetual motion,
they are as magical as aurora—
orange gold yellow-green red iridescent,
glowing as they hum in our midnight sun.

Mountain Goat

Oreamnos americanus

On rock faces that appear entirely vertical, high in jagged peaks where mountains dive into the sea, Alaska's Mountain Goats stand calmly, safe in their spring home. Like ghostly apparitions in their long cream-colored coats looking down from above, Mountain Goats animate the steep alpine world into which humans seldom venture.

Thriving at altitudes up to 13,000 feet in the Wrangells, they're so often seen along one section of coastline in Kenai Fjords National Park that it's referred to as goat alley. But they're not true goats; they're actually the single North American representative of a unique group of mountain ungulates called rock goats. They're more closely related to gazelles and antelopes than to true goats.

Unlike Dall sheep, who live in mountains farther inland and have curling tan horns that grow throughout the males' lives, Mountain Goats have shiny black horns that grow 8 to 12 inches long in both males and females. They're also larger, with stocky males reaching 300 or more pounds. Their narrow rib cages, muscular shoulders, and unique hooves allow them to walk thin ledges. Their hooves consist of two toes that move independently from each other, each with a hard outer sheath that gives them purchase in cracks, and soft, spongy foot pads that can grip wet rock. They are the ultimate mountaineers, able to navigate seemingly impossible cliff faces.

Their astonishing agility on the steepest high-altitude alpine and subalpine terrain keeps Mountain Goats safe from predators, so much so that the main cause of their deaths, outside of human hunters, is avalanches. True herbivores, they eat a diversity of plants, including mosses, ferns, lichens, grasses, shrubs, and tree buds. After a long undernourished winter, some bands move to the shoreline to graze on kelp and then slowly follow the snowmelt up mountainsides, grazing as summertime growth appears. Like bears, they pack on fat in late summer to survive winter.

In Alaska, Mountain Goats thrive along the coasts from Southeast into the Kenai Peninsula and in the Kodiak Archipelago. Other populations range farther inland, as far as the Talkeetna Mountains, where they overlap with Dall sheep.

Mountain Goats are particularly sensitive to human disturbances, especially helicopter tourism and industrial mining. Timber harvests reduce their habitat, and overhunting—especially of females—can cause quick drops in population.

They're also highly at risk from changing winter conditions brought on by climate change, such as higher snow levels, more freeze-thaw cycles, increased avalanche frequency or intensity, and raging winter storms.

Little is known about the lives of Mountain Goats, however, because they live in such difficult-to-reach terrain. What we do know is that in some of the most spectacular alpine landscapes on Earth, Mountain Goats live on the edge.

HEATHER LENDE

Jánwu Jeté

In memory of Lingít weaver Teri Rofkar, 1956–2016

Fish & Game slides
of Sitka and Chilkat ridges—
nannies and kids circle
and prance

on peaks so slick and narrow
the cameraman needs an airplane.
Impressive,
he whispers.

Imagine living on
            lichen
and faith
            in your
                        feet.

A billy delights
in fancy white pantaloons
fearless, gallant,
impossibly
            balanced.

The weaver
presents a dancing robe

its pattern binary code,
goat wool strands spun
hand on thigh
in Lingít time.

He's got his beautiful
presentation.
She says
we both collect the same
            information.

She swings the fringe.
See these tassels.
Feel the motion.
            This robe wants to dance.

They twirl and prance,
fearless, gallant,
perfectly balanced.

Nunatak

Ragged and sharp, like a dragon's back or giant horns, they appear mirage-like above smooth white sheens of ice. From the Greenland Inuit word *nunataq*, meaning "lonely peak," *nunatak* is the term for these ridges or peaks that rise above a glacier or icefield. These craggy, stark silhouettes towering over the surrounding landscape testify by their very shapes that they have never been ice covered, have never been subject to the grinding, rounding force of great sheets of ice.

Remaining above the ice during the last glacial maximum approximately twenty thousand years ago, Nunataks now provide clues to that last ice age. Periglacial trimlines along their flanks, with their smoothed rock and scree slopes, mark the upper level of glacial erosion and reveal just how high glacial ice once reached.

Seemingly formidable and lifeless from the great distances across which most of us see them, Nunataks can harbor a surprising diversity of life. When they defied the great ice sheets, these glacial islands became refugia. In Alaska, some of the plants who escaped ice this way are also found in the high Rocky Mountains and on the Pacific coast, indicating their once-wider range. Other plants found on Nunataks are endemic to Alaska and the Yukon. Besides harboring rare plants and a few hardy insects, Nunataks—with their geographic isolation and sensitivity to environmental change—may make it possible for species to reestablish in recently deglaciated areas.

As the glaciers and icefields around them shrink and recede, some Nunataks are being reclassified as just mountains. Or they may be called Paleonunataks, a term used for summits formerly surrounded by ice. Or Arêtes—thin, saw-toothed crests that once separated two glaciers. Or, for those with a pointed peak and sharp, distinct edges like a pyramid, Horns. Great Nunatak, a mountain in Southcentral Alaska's Chugach Mountains, was so named in 1899 by G. K. Gilbert, a geologist with the Harriman expedition. The glacial lobes that encircled this once-Nunatak have shrunk as the Columbia Glacier has retreated, and open ocean now replaces ice on two sides of what is now called, simply, a mountain.

Though few people, save those who traverse icefields, ever see them up close, Nunataks can be seen from afar throughout Alaska, wherever glaciers and icefields spread, particularly along the coastal mountain ranges of Kenai Fjords, Lake Clark, the Chugach Mountains, the Wrangells, and the Tongass. With their serrated, fractured forms, these unique geologic features spark imagination even as they reveal the powerful forces of the planet.

JEREMY PATAKY

Thin Air Tautology

Lichen plinth, rock pearl, era mote,
you conduct ice arias in remote halls of winds.

Everything flows toward you, past, and away.
The sun laves you with light that you tongue into shadow.

You erupt from the body of ice, ice of unknown depth.
Your self a compound fracture. Original sundial
spreading rumors of an alleged valley floor.

Bones (animals, your own) you grind like coffee
to stir into ice tides

that flow and ebb, but always go,
marking you as you in turn mark.

Winter bakes the color black into your teeth.
You throb stillness and sweat weather
and stop your own breath
for human lifetimes
as the scintillant, chaffing fetch complicates.

Low-slung trumpeter swans migrate forever by,
little heartbeats pumping wings there and back
until they mutate into overflying jets—

one skitters its own mechanical shadow
over the glacier's ornate hide,
eclipsing, for a blip, a peregrine.

Humans inside the jet have silt grain pupils
and they look at you, or don't,
through thin air over the range you're of—

then, more pure quiet.

Who has known the views as you do?
Goat, sheep, wolverine, falcon, woman? And lichens,
clotted synapses bred by golden hours?

Marooned building block fumbled in the frenzied heat of creation—
Compound eye regarding square miles of ice all around,
and deep space, heavens, satellites,
and all the gothic weathers:
volcanic ash from eastern Asia.

Saharan dust from a storm in Africa.
Seabirds struck stiff some inexplicable miles from the sea.

The last law of nature is a tautology:
Everything in the world gets a witness.

Even your flanks, finally, hidden until the
little ice age made its left stage exit.

Even the first dragonfly to traverse
glacier airspace to light on you in anonymous alpenglow.

Even the spiders who dust your crevices
with their own bodies.

Western Hemlock

Tsuga heterophylla

Sitka Spruce

Picea sitchensis

Towering together in deep forests that hug the coastline, a thin green membrane between ocean and ice-covered mountains, Sitka Spruce and Western Hemlock anchor the great temperate rainforest that extends up Alaska's southern coastline to Prince William Sound, coastal Kenai Peninsula, and the Kodiak Archipelago. As the dominant climax trees in Alaska's coastal forests, these twin giants provide an exceptionally diverse habitat for a host of wild beings.

Here, Sitka black-tailed deer, lichen, salmon, and peregrine falcons can be found. Hairy woodpeckers tap old trunks to find insects and create nest cavities. Porcupines and black bears climb their thick trunks, finding safe harbor in their sheltering branches. Deep cushions of sphagnum moss thrive beneath them, holding up to forty times their weight in water.

These big trees can live long lives. While much of the old-growth forest was logged before we knew how to determine tree ages, it's estimated that Sitka Spruce—Alaska's state tree—can reach eight hundred years old, and Western Hemlock, twelve hundred. And they're huge: both can grow to more than 200 feet tall, with Sitka Spruce trunk diameters exceeding 8 feet. Sitka Spruce is by far the largest species of spruce and the third-tallest conifer in the world.

These giants largely grow on thin, poorly drained soil and are susceptible to windthrow. Most openings created this way are small and provide space for young trees to grow, enhancing the durability of natural mixed-age forests. Along steep shoreline cliffs, the long lateral roots of Western Hemlock and Sitka Spruce can reach into deep cracks for stability. When these trees do fall and rot, they contribute to healthy mature forests, providing nutrients and shelter to many beings.

Needles provide an easy way to distinguish the two species. Sitka Spruce needles are sharp and tough, and their branches push out to an erect crown tip. Western Hemlock has soft, flat, feathery sprays of needles, and their crowns end in a graceful downward curve.

It's harder to distinguish Western from their Alaska cousin mountain hemlock, except by where they grow; mountain hemlock is found on peatlands or colder, subalpine sites, where winds often shape them into bonsai-like forms called krummholz. And Sitka Spruce is much larger than white or black spruce, who grow farther inland.

Sitka Spruce ranges farther than Western Hemlock, draping the northern and eastern parts of the Kodiak Archipelago with thick, dense forests. There are even small stands of Sitka Spruce in the windy, treeless Aleutians. These groves were all planted by homesick humans. One grove, planted by Russians in 1805, is considered the oldest recorded afforestation, or forest planting, project in North America.

Both trees have long been used by Alaska's Indigenous Peoples for totem poles, canoes, and houses, as well as traditional tools like arrows, spear shafts, fish drying racks, and snowshoes. Roots are twined into watertight hats and baskets, ropes, and fishing lines. Spruce pitch has been used as caulking, chewing gum, and in traditional medicine. In the spring, Hemlock branches are submerged in inshore waters to collect herring eggs during the spawn.

For many years, the forests of Southeast Alaska were clear-cut, their trees turned into pulp, which was then made into rayon fabric and paper. Today, Sitka Spruce, highly valued for strength, elasticity, and tonal quality, is the standard for guitar soundboards (tops) and is a select wood for other fine instruments and woodworking.

Only remnants of old-growth forests can still be found in Alaska, and the absence of that mixed-maturity forest makes existing trees more vulnerable to insect damage. Insect infestations are also increasing with a warming climate, and changes in rainfall and snow patterns make the two giants even more susceptible.

Deep within forests of mature Sitka Spruce and Western Hemlock, quiet reigns. With soft moss underfoot and massive trees sheltering overhead, someone walking there on the frequent rainy days can easily hear even the smallest bit of water fall from a branch overhead onto the wide umbrella leaf of devil's club; the sound is like laughter, or a song.

THOMAS R. BACON

In the Shade of Hemlock and Spruce

Cloud or fog
the white breath of ancestors
crosses to the far bank
in the shade of hemlock and spruce.

Twilight peels sunlight from clear water,
currents of shadows bridge the river,

and a lost breeze dances into memory.
At nightfall, moss carpet to dreams,
we rediscover our first moonbeam,
remember ancient patterns of stars.

Common Raven

Corvus corax

Ravens are ubiquitous in Alaska, living in habitats from dense forest to tundra. They follow hunters and wolves to food sources, sit on dumpsters and light poles in urban areas, and generally make themselves known with their bold presence. Their croaking and knocking voices punctuate winter silences, and their distinctive soaring flight, playful acrobatic rolls, and somersaults amuse humans

as well as themselves. They drop and catch objects in midair and sometimes slide down snowfields just for fun.

Ravens are among the most imposing of Alaska's birds. Physically, Ravens resemble their smaller American crow cousins but are distinguished by their thick bills and shaggy neck feathers. The largest of what are classified as songbirds, Alaska's Ravens can have wingspans of up to 5 feet and weigh more than 3 pounds. They're completely black in color—not just their feathers but also their legs, bills, and eyes, black as an Alaskan winter night.

Ravens build massive stick nests in the crotches of trees or, in treeless areas, on cliffs beneath an overhang. The females lay three to seven eggs, and the chicks leave their nests five to six weeks after hatching. Mated pairs stay together throughout the year and for their lifetimes, which can reach more than twenty years.

Ravens are more solitary than crows, but young Ravens typically stick together in groups. In winter, Ravens form loose flocks. As with most birds who winter in Alaska, Ravens pair up and spread out in summer for nesting and chick rearing but in winter flock together for safety from predators, for warmth in roosting together, and for sharing information about food sources. On winter days, they travel up to 40 miles from nighttime roosts to daytime feeding places. They'll follow a similar route every evening, talking loudly as they fly, to gather in the same large tree, often a cottonwood. These roosting congregations are far from human habitation and can be very large: as many as eight hundred have been counted at one roost in the Interior.

Like all corvids, Ravens are incredibly curious and intelligent—they're known to be among the smartest of birds. They talk with large vocabularies, using at least thirty-three different sounds that are discernible by humans, and can mimic other birds and even human speech. They use tools, and they engage in a complex and cooperative social life. They understand cause and effect, and will follow the sound of gunshot to find freshly killed food while ignoring other equally loud noises. They problem solve. In one test, a piece of meat was tied to a string hung from a perch. Ravens figured out how to lift the string with their beaks, step on the string to hold it in place, lift another section, step, and finally reach the prize. They can even recognize individuals—other birds and animals including humans—by sight. Researchers who capture Ravens found that the birds identified and avoided them; those researchers did their own problem-solving by changing clothing and wearing masks.

Throughout the world, Ravens hold a prominent place in mythology, folklore, art, oral history, and literature. Alaska's Indigenous Peoples have long admired the Raven's intelligence; in their stories, Raven is a frequent and respected character with supernatural powers. In Tlingit and Haida cultures, Raven is both a trickster and a creator who brought daylight to the world. Tlingit, Haida, and Eyak divide into either Raven or Eagle moieties.

A rare white Raven frequenting the Spenard neighborhood of Anchorage in the winter of 2023–24 became an internet celebrity with her own paparazzi. This bird was not albino but leucistic—lacking pigments besides melanin—and had blue rather than black eyes. A white Raven in many mythologies is a messenger bird with special, god-adjacent significance. In the Haida and Tlingit cultures, Raven was originally white and became black.

Raven numbers in Alaska and elsewhere are on the increase, since human activities—producing crops, garbage, and roadkill—provide them with food. They will eat almost anything, including smaller birds, bird eggs, insects, rodents, pet food and feces, grains, berries, and carrion. Their predation on other birds can reduce those populations.

Unlike most wild beings, Ravens readily interact with humans, in ways both delightful and pesky. They'll talk to us and approach us, but they'll also steal our food, peck holes in airplane fabrics, and fly off with shiny objects and golf balls. For a while on the Kenai Peninsula, some Ravens had a habit of stealing windshield wipers. One video shows Alaska's white Raven loosening a bolt on a streetlamp and carrying the bolt away in her beak. Intelligent and inquisitive, Ravens are welcoming companions, especially in winter, when so many other birds have migrated south and these sleek black beings gather together in big, noisy, magnificent flocks.

ROBERT DAVIS HOFFMANN

Raven Moves

If I make words, they are Raven's echo.
If I move, it is in that rhythm, Raven's heart,
the air pulsing and rumbling.
It makes me restless.

In the back of my mind,
long-ago night ritual
men crouched in a circle,
drumming the earth.

Raven gets excited,
darts here and there
gathering leaves and bone,
tying all things together
into this dark nest.

Raven belly swells,
he lifts his wings and buries his head.
The night swirls into itself
and I emerge,
blinking,
a dull thudding in my ears.

Alaska Monkshood

Aconitum delphiniifolium

In midsummer, as cranesbill geranium, wild rose, and lupine fade, the deep purple blooms of Alaska Monkshood appear. Along with yarrow and fireweed, these flowers linger into fall, blazing violet as birch leaves turn gold, grass blades tip burgundy, and red berries brighten carpets of Canadian dogwood.

Thriving throughout most of Alaska, in forests as well as subalpine meadows and moist swales in high alpine tundra, clusters of Alaska Monkshood flowers rise up on slender stalks with deeply lobed leaves. Their height depends on who grows around them and ranges from 6 inches to 4 feet. The other species of *Aconitum* in Alaska, Kamchatka aconite (*Aconitum maximum*), towers to 8 feet tall in the rich coastal meadows and thickets of the Aleutian Islands and Southwest Alaska. Worldwide, some *Aconitum* species are grown as garden plants.

It's the unusual shape of the flowers—the top sepal shaped like a hood covering the nectary—that gives them their common name. The genus name is derived from the Greek *akoniton*, which translates to "wolfbane," their other most common name, from their ancient use in Europe to poison wolves by mixing roots with meat scraps.

Indeed, Monkshood is highly poisonous when ingested and is the most poisonous plant native to North America. The most toxic part is their tuberous roots; the Alutiiq used spears dipped in the ground root to hunt humpback and minke whales. Worldwide, other *Aconitum* species have a long history of use both as poison and medicine, and make appearances in Greek myths, Shakespeare's *Henry IV*, and *Harry Potter*.

Fortunately, the poison doesn't bother Alaska Monkshood's pollinators. Long-tongued bumblebees have the strength to open the hooded flowers and reach the single nectary inside, the sweet secret only they can divine.

JOAN NAVIYUK KANE

Hyperboreal

Arnica nods heavy-headed on the bruised slope.
Peaks recede in all directions, in heat-haze,
Evening in my recollection.

The shield at my throat ornamental and worse,
We descended the gully thrummed into confusion
With the last snowmelt a tricklet into mud, ulterior—

One wolfbane bloom, iodine-hued, rising on its stalk
Into the luster of air: June really isn't June anymore,
Is it? A glacier's heart of milk loosed from a thousand

Summer days in extravagant succession,
From the back of my tongue, dexterous and sinister.

Autumnal Moth

Epirrita autumnata

The Autumnal Moth is a delicate moth with a wingspan of just over an inch. Colors vary from gray to white, with paisley patterns of darker gray on symmetrical wings. Their range in North America reaches from the northern states through southern Canada and into Alaska, where the species was only

identified for the first time in 2005. Autumnals are found in Eurasia at similar latitudes.

The caterpillar, a green inchworm known as a green velvet looper, feeds on the leaves of alder, willow, birch, and poplar trees as well as berry bushes and some conifers. In big years, the caterpillars in Lapland are said to defoliate entire mountainsides of birch trees. Those big years come in ten- or fifteen-year cycles; environmental factors causing the cycles are not well understood but likely correlate to temperatures. In 2011, a news article reported huge numbers of the moths "mobbing" Southcentral Alaska during a cycle peak.

Eggs are laid in the fall and overwinter, and the caterpillars feed from May to July before burrowing into the soil to metamorphose into moth form. The moths fly in September and October (hence their common name) and then die with the colder weather. The eggs, laid in tree bark crevices, are said to survive temperatures as low as –50 degrees F.

Alaska has 127 known moth species and about 80 known butterfly species. Moths and butterflies are not always easy to tell apart. The best way is to get a close look at the antennae: butterfly antennae have little knobs on their ends, while moth antennae are plain or feathered.

Like other moths, Autumnal Moths are drawn to light. A good place to find them in the fall is around porch or bike path lights. Otherwise, they're found all summer long flitting through the woods among the green leaves they love.

MAR KA

Eclipse of Autumnal Moths

—for Sijo

The morning after a full moon, we come,
my daughter and I, of a sudden, upon,
frozen into the now thinly-iced pond,
what looks to be a cherry blossom path,
what looks to be a fairy blossom path,
tiny, papery, white petals strewn long.

It's past the time of blossoms. Past the time
of berries. And here above the tree line
near the top of Chugach State Park's
Baldy Mountain, we're beyond the reach
of destination-wedding flower baskets.
Mystified, entranced, we kneel, glad

for insulating pants, at the edge of the ice;
use the lenses on our iPhones to magnify
the unexplained phenomenon—see flight!
The petals are wings! These are moth bodies!
Flown into the reflected full-moon light
just as last night cooled and the pond iced.

Hundreds and hundreds, flat-pressed
between ice-plates, like specimens
or keepsakes. Our fingers flutter like moth

wings on our camera shutters. Ping! Ping! Ping! Ping! Ping! Ping! White feathery things against dark water. Marvel. Metaphor.

Devil's Club

Oplopanax horridus (synonym *Echinopanax horridus*)

The Latin name says it: *Oplo* translates to "weapon," *panax* means "healer-of-all," and *horridus* means "horrible." The Dena'ina name Heshkeghka'a translates to "prickle big-big." Devil's Club, also called Devil's Walking Stick, is both someone to avoid and someone to seek out as a healer.

This deciduous shrub grows in wet wooded areas, in nitrogen-rich soils and often impenetrable thickets—impenetrable because of the sharp spines on the stems and the undersides of leaves. The towering plants, who can reach 10 feet tall, are quite beautiful, with their lush green leaves that turn golden in fall and their torches of red berries, inedible for people but favored by bears, squirrels, and thrushes. One study found Devil's Club to be the most important summer and fall food on the Kenai Peninsula for black bears. Any black bear scat in fall will corroborate: it's frequently packed full of bright red Devil's Club seeds.

These plants, who thrive in Southcentral and Southeast Alaska's forests, are vital members of their wild community. Their broad leaves provide shade and cover over salmon streams, and birds and small mammals find protection among the spiky stems. And while the clusters of white-green flowers in a pyramid shape seem inconspicuous to us, outside of their intoxicating ginseng scent, pollinating insects find them very attractive. When black bears spread their seeds through droppings, they help forests to recover from disturbances, including landslides and logging. The armed-to-the-teeth plants also serve an ecological purpose by forming a natural barrier around fragile wetlands.

A member of the ginseng family, Devil's Club has been used as medicine for any array of ailments including colds, depression, and broken bones. "Healer-of-all" indeed: Devil's Club is considered to be the most important ethnobotanical in Alaska and the Pacific Northwest. The bark and root are the harvested parts. Extracts from Devil's Club are marketed for a variety of ailments, while

medical researchers continue to study the plant's phytochemicals, including for anticancer activity.

Besides serving as food and medicine, Devil's Club has been employed for spiritual and practical purposes. Plant parts have been used in purification rites and to ward off evil, for making face paints and dyes, and for fashioning fishing lures. Because Devil's Club is considered a sacred plant, it is best when gathering to learn from and accompany a traditional medicine harvester. Moreover, except when purposely gathering, it's wise to avoid the spiny stalks and leaves, the thorns of which can embed and fester in flesh for days.

Why does Devil's Club go to such lengths to ward off others? Biologists think the protective armor guards the leaves, which were likely much sought after in the plant's evolutionary past because they are so nutritious. Aside from slugs, members of the animal world now largely leave the leaves alone. Still, it's a springtime delight to see a thicket of those wicked-looking thorny stems burst with bright green fists, ready to unfurl their sheltering leaves.

LINDA MARTIN

harboring a mean streak

The patch of devil's club in my yard
lifts fiery spires of poison berries
on spiny stems above thorny leaves.

Vicious native plant. Touch it, you'll be sorry.
Even the berries have stickers.

Were I to dig it up, the soil would prove fertile.
Other shrubs with nicer traits could thrive there.
But what a chore, the roots are deep,
and I would miss that prickly wildness.

Brown Bear

Ursus arctos

Perhaps no being is more iconic for Alaska than the Brown Bear, who can measure up to 10 feet high when upright. This imposing mammal stands tall in the Indigenous cultures of Alaska, as well as on Alaska's license plates, Alaska's commemorative quarter, and in innumerable photographs, works of art, and pieces of visitor memorabilia.

Brown and Grizzly Bears are classified as the same species, although some scientists break *Ursus arctos* into three subspecies—Coastal Brown Bear, Grizzly Bear, and Kodiak Brown Bear. In Alaska, the name "Brown" is applied to those who live along the coasts, where they feed on seasonally abundant salmon plus rich vegetation; they grow larger and live in higher densities than the Grizzlies who live in the interior and northern parts of the state. Both bears feature a lot of color variations, from pale blond (most common among Grizzlies) to dark brown, so color alone cannot distinguish them from black bears.

Although only about fifteen hundred Brown Bears survive today in the lower forty-eight states, the approximately thirty thousand in Alaska frequent most regions except for islands in the south and far west. The population on Admiralty Island, in Southeast Alaska, at an estimated sixteen hundred, outnumbers the island's human population by more than two to one.

In Alaska, Indigenous stories, beliefs, and practices show a close relationship and respect for this physically and spiritually powerful being. The Tlingit story "The Woman Who Married a Bear" indicates the familial relationship between Brown Bears and humans. The Koyukon Athabaskans traditionally showed respect by referring to these bears only by circumlocution: as "the big one" or "Grandfather."

Most Brown Bears spend winters in dens, where they do not eat or drink, urinate or defecate. Instead, they recycle metabolic waste and rely on stored fat for energy. This is not true hibernation, only a way to save energy by reducing their body temperature and heart rate.

Between one and three cubs are born in a den during January or February and emerge in June with their mothers, at which time they weigh only a few pounds. The cubs require two or three years with their mothers before they have the size and ability to survive on their own.

When the bears emerge in late spring, they will have lost up to a third of their body weight. So it's not surprising that they're ready to eat just about anything,

including new plant shoots, sedges, roots, clams, salmon, berries, small mammals, and sometimes the young of large animals like moose, caribou, deer, and musk ox. By fall, Brown Bears can double their weight if food availability allows. At peak condition in the fall, a Coastal Brown Bear can weigh 1,200 or more pounds.

In fact, prior to winter denning Brown Bears enter a period of insatiable eating and drinking—called hyperphagia—to put on fat for winter. Hyperphagia is triggered by the suppression of leptin, the hormone that signals satiation, allowing them to continue gorging themselves until hibernation. During this period, Coastal Brown Bears can consume up to thirty salmon and put on 4 pounds each day. In late fall, leptin suppression ends, signaling that it's time to stop eating and sleep. This remarkable adaptation allows Brown Bears to take full advantage of Alaska's summer abundance and then snooze through the long, dark winter.

Generally solitary or in mother-and-cub groups, Brown Bears gather where food is concentrated, like streams during salmon runs. Here they develop complex hierarchical social structures to avoid conflicts. These gatherings also draw human wildlife watchers. At several places along the Alaska Peninsula and in Southeast, viewers learn to distinguish individual bears by unique physical characteristics as well as their personalities and fishing styles. A live camera at Brooks Falls in Katmai National Park allows anyone anywhere to watch Bears fishing and interacting. Fat Bear Week is an annual fall celebration that lets the public vote on the fattest bear who's been seen on camera.

A group of bears is called a sloth, a word assigned in medieval times because European hunters apparently thought they were slow—but nothing could be further from the truth. When motivated, a Brown Bear can sprint as much as 30 miles per hour.

Brown Bears have an acute sense of smell, even sharper than that of dogs. Contrary to popular belief, their eyesight is also very good, comparable to that of humans. When a Brown Bear stands upright, it's not a sign of aggression—only an effort to get a better look at you.

DAVID MCELROY

Bear 747

sits in the river's Jacuzzi below the falls
waiting for the bump against his legs
that is a salmon roiling blind in bubbles.

We stare at 900 pounds of not exactly troll,
appetite of boar, the power of bear
and the falls where he stares.

A quick plunge of his head and up comes
a red flopping in his mouth. Motor-drive
cameras, each lens big as your leg, track

the head shake, water spray, and turn, his back
to the viewers, his skill at stripping skin,
meat, and eggs, feeding these tiny steps

to winter survival. Gulls gather
for scraps, and magpies peck scat
for tapeworms. "They're not human

so we give them numbers," a ranger
informs with the gravitas science inspires.
But the four-year-old sub-adult 284

splashing in the shallows, or like a hippo
snorkeling the lower river, climbing trees,
yawning, false charging through camp,

popping her jaws is something else. Raised arms
and loud talk make us big, our numbers bigger,
we think she thinks, so human we are with fate or fear,

and we call her Wild Child. The emergency call
that comes like the bear you don't see, shape shifting
in the night—thick brush, wrong number on a kill—

comes for our friend. A float pilot, who crashed
the week before and survived, arrives to fly her off
the water's wake and into the ache of air.

On bigger and bigger planes she flies
to her son who lies in a coma trussed
in the elegant machinery false hope requires.

Bench pressing 350 pounds,
something happened, no name for it—
his number up—call it Freak Accident.

The bar came down and pinned his chest
causing him to regurgitate and aspirate
his breakfast, the peach that killed him.

Moraine

Like the words for so much that has to do with glaciers (*cirque*, *arête*, *crevasse*, *moulin*, *serac*, and *glacier* itself), the word *moraine* comes from the French. It means a "ridge of rock deposited along the edge of a glacier." The term was introduced into geology in 1779.

There are many types of Moraines—the most familiar being lateral (adjacent to valley walls), medial (a ridge away from walls, on a glacier's surface or in a valley, and formed by the joining of two lateral Moraines), and terminal (marking the farthest point reached by an advancing glacier). Other types are ablation, ground, ice-cored, push, and recessional.

In every case these ridges, also known as glacial till, are made up of loose rock debris scraped and pushed by glaciers. Such material can vary in size from large boulders to gravel to finely ground dust referred to as rock flour. Moraines exist not just around today's glaciers but also in areas previously covered by glaciers.

Alaska's famous Homer Spit is one example of a terminal Moraine, left by a glacier that once filled Kachemak Bay. In Juneau, Moraine Edge Subdivision includes a prominent ridge that marks where the terminus of the Mendenhall Glacier sat as recently as 1760.

Moraines and other landforms produced by glaciers and their movements record the nature and timing of past climate changes and can help us understand what we might expect in the future. In Denali National Park, Moraines are used as a standard by which to compare the timing of glaciation in other parts of Alaska. The technique of cosmogenic radionuclide dating can be used to obtain ages of Moraines formed more than a million years ago. From the amount of a particular radionuclide in a sample, scientists can calculate how long a rock has been sitting on Earth's surface.

Some Moraines are easily noticed, such as the medial type, showing up as dark streaks like highways pouring down white rivers of ice, while others are less obvious. Other evidence of glacial activity includes glacial erratics, massive boulders dropped by retreating glaciers and now hunkered deep in a forest, draped in moss and lichen. Moraines and erratics found throughout Alaska's landscapes remind us of the ever-shifting nature of this northern land.

SHEHLA ANJUM

An Absence Regained

With silvery snow and azure ice,
With sculptural seracs
and labyrinth of crevasse,
A glacier awes and delights.
But marring its splendor,
behold heaps of dirt and stone—
the lateral, medial, and terminal moraines—
arising from a glacier's ceaseless
pulverizing of mountain and valley
to boulder, rock, and till.

When a glacier, like all beings,
ceases to exist
the moraine's presence
will fill its absence.

Ridges rising along mountainsides
will tell a glacier's height.
The contours of a valley
will recount the story
of its silent flow.
Mounds, eskers, kettles,
washboards and drumlins
will shape the land
as testament
to the glacier's icy might.

In valleys where ice did once reign
these echoes of past remain—
haunting reminders of a bygone grace
they usher in a new terrain.

Ruby-crowned Kinglet

Corthylio calendula (formerly *Regulus calendula*)

Flitting among branches, never still for more than a nanosecond, the Ruby-crowned Kinglet is a tiny bird, about the size of a hummingbird but more rounded. More than the Golden-crowned Kinglet, a somewhat-less-common relative, the Ruby-crowned is exceedingly active, acrobating through Alaska's mature spruce forest, sometimes hovering to snap at insects. An attentive human is more likely to hear their song—bubbly and surprisingly loud and complex, with many variations—than to spot one, although the white eye-ring and especially the constant wing flicking are giveaways for identification. It's the male who wears the red crown, which usually lies flat and thus is hard to see but which will stand in a blaze for courting a mate or challenging a competitor.

The Ruby-crowned Kinglet's manic activity level results in a high metabolic rate—a heart rate of 1,200 beats per minute (and still half that when resting). While these birds feed mainly on insects and spiders, they will also eat berries and tree sap. They can lose 10 percent of their body weight sleeping overnight—and gain it back the next day.

This tiny ball of feathers lays a large clutch of eggs—up to a dozen—high in a tree. The elaborate and beautiful nest resembles a hanging cup, built of moss, twigs, lichens, and spiderwebs and lined with feathers and animal hair. Males don't help with the nest or eggs but do stay with the females and help feed their chicks until they fledge. By October, parents and young head south, leaving quiet woods behind them. They usually live for about five years, although one who was banded was seen for more than eight years.

Because Ruby-crowned Kinglets breed mostly in remote forested areas in the north and winter in a variety of habitats throughout North America where they tolerate landscape disturbances by humans, these royal birds are of low conservation concern. Numbers rise and fall regionally, depending on winter weather and food availability. Logging and wildfire are threats to their forests and lives. Climate heating may influence their migration timing or range.

In many parts of Alaska, the beautiful, resounding song of the Ruby-crowned Kinglet—soft, high notes that shift into louder repeated notes, all lasting about five seconds—announces the commencement of springtime, with all the diversity of birdsong that it brings.

KIM HEACOX

A Regal Bird, Small But Mighty

> We have lived our lives by the assumption that what was good for us would be good for the world. We have been wrong. We must change our lives so that it will be possible to live by the contrary assumption, that what is good for the world will be good for us. And that requires that we make the effort to know the world and learn what is good for it.
>
> —Wendell Berry

Every April, like clockwork, I hear my first ruby-crowned kinglet,
and the world, despite its problems, is beautiful again.
My heart gladdens.

It's more than a sound. Or a song.
It's a proclamation: a regal bird, small but mighty,
saying, "Yes, I'm back.
And I aim to make the most of it."

The willows have barely begun to leaf out. Nighttime temperatures
still fall below freezing. Snow squalls—the last gasp
of a long winter—come from nowhere. Yet the kinglets arrive
from wintering grounds thousands of miles to the south.

Once back north and no doubt exhausted, do the kinglets rest?
They must. But I only ever see them in constant motion,
flitting among the branches, flicking their wings and hanging upside down,
determined to feast on what Richard Nelson called
"the great Alaska bug factory."

How adroitly they snap mosquitos, dip them in tree sap
(if a red-breasted sapsucker has been busy nearby),
and hurry off to feed chicks who wait in cup-shaped nests of moss,
perhaps forty feet up a sheltering Sitka spruce.

When I hear my first kinglets
I allow myself to hope for the best.
And promise again to learn what is good.

Alder

Alnus alnobetula (synonym *Alnus viridis*)

Hikers in Alaska are not fond of Alders, who grow in dense thickets where they extend their trunks and branches horizontally, the better to block a person's passage. Bears and moose, however, have no trouble getting through. Though they may be cursed by humans, Alders, along with devil's club, help protect wild areas by discouraging entry. They're also essential inhabitants of their neighborhoods, where they thrive in poor soils. The nitrogen-fixing nodules on their roots prepare the soil for other plants to follow.

Of the four Alder species found in Alaska, Sitka (*Alnus alnobetula* subsp. *sinuata*) is the most common, inhabiting most of the southern half of the state. As

deciduous shrubs or small trees, Sitka Alders (Uqgwik in the Alutiiq language, Qenq'ena in Dena'ina Athabascan, Keishísh in Tlingit) grow primarily along shorelines and streams, in recently deglaciated areas, in avalanche chutes and mountain swales, and in clearings made by humans. A pioneer species, they're among the first plants to populate disturbed places and grow quickly. They can grow to 20 or more feet tall, although they are typically under 10 feet and bushy. They require sunlight, so once spruce and other trees grow up around them, Sitka Alders are squeezed out. They do well in avalanche tracks because snow slides over the sprawling branches without breaking them.

The Alder's leaves are shiny green, with saw-toothed edges. The bark is smooth and gray. Male and female flowers appear as catkins on the same plant. The female catkins are wind-pollinated and become cone-like fruits at maturity. The green cones turn brown and generally hang on through winter.

Moose usually find Alders unpalatable, but beavers, muskrats, and hares eat their leaves and twigs, and beavers build lodges and dams with them. The seeds, buds, and catkins are important food sources for many birds, including chickadees, redpolls, siskins, and warblers. Alder's dense thickets provide essential cover for large and small beings, from moose to songbirds. The parasitic plant northern groundcone, who is a favorite food of bears, draws nutrition from Alder's roots. Other beings that associate with them are willows and ferns; lichens often grow on the bark.

The bark and leaves of Alder are very astringent, and Alaska's Indigenous Peoples have long used them medicinally. A tea made from boiling the green fruits treats various ailments including diarrhea. Leaf poultices ease the pain and itching of insect stings and bites. Leafy branches used as switches in steam baths and saunas are said to relieve aches and pains and promote good health generally.

Alder can be burned as firewood, but it's particularly valued for smoking salmon. Because of the wood's flexibility, it is used by Alutiiq (and others) to build snowshoes, kayaks, and fish traps. Alder is also a preferred wood for carving dishes and utensils, since it doesn't transfer taste to food. Alder bark, when soaked or boiled, makes a reddish-brown dye.

Although human hikers may find the tangling growth of this plant annoying, there's no doubt that the Alder is a generous being. Alders stabilize stream banks, providing shade for young salmon. With their flexible limbs and soil-building roots, and the junglelike refuges they create, Alders protect and sustain life everywhere they grow.

TARA BALLARD

Magic Maybe

I ease myself in and through the branches,
longer and higher than I will ever grow.
Lime bright with sun, new leaves stick
to my fingers as I measure the steps
toward retrieving our soccer ball, tucked
now at the base of trees. Our game paused
by the fact that one goal is marked
by the tangle that separates our home
from neighbors. Where she waits
on the grass, my sister calls to me, offering
a child's directions to another: If I look back,
I would see that she points to the left.
That way. But I already know. Everything
seems to find rest in the same location.
The far corner, nudged between a V
of growth, as if settled there on purpose.
This stretch of alders, in November, breaks
our speed as we race across the snow
in matching sleds. Come June, the bush-like trees
provide an imagined fort, a secret hideaway,
magnet for every sport we play. If I dig
deep enough, I will find the birdie
to our badminton and at least one frisbee,
forgotten two seasons prior. As I near
the ball, I slide my thumb and finger
down a trio of heavy catkins, dangling
like grapes. Golden pollen slips earthward
as I am certain fairy dust would. In the alders,
I think, so much is possible: The deeper
I wade, the quieter. The greener I turn.
Our soccer ball becomes a nestled egg.
Half expecting a glow, I reach for it
with both hands. My sister's voice changes
to robin song, far off and light.

Black-billed Magpie

Pica hudsonia

Tiding, *conventicle*, *colony*, *gulp*, *mischief*, or *tittering*: a flock of magpies goes by many names, but the most telling in terms of how we see them might just be *mischief*.

A mischief of magpies usually consists of family members; they're gregarious and talkative, with calls sounding like people chattering. They flock together in winter, sharing food sources and roosting together at night. In Alaska, they're sometimes seen above the Arctic Circle at Fort Yukon, and throughout their range prefer mixed forests and open areas.

Strikingly patterned black-and-white birds with long tail feathers, Black-billed Magpies are large, with a 2-foot wingspan. Their iridescent black plumage shimmers green and blue and purple in sunlight. They mate for life, building huge, elaborate, dome-shaped nests of woven sticks with two entrances and a mud cup, high in cottonwood or birch trees. They tend their young for six months, much longer than most birds. Sometimes fledglings—still not quite feathered and looking very prehistoric—will fall to the ground, and the parents will continue feeding and noisily protecting them for days, until they take flight.

Members of the corvid family, magpies have that family's reputation for inquisitiveness and intelligence. Considered intelligent by the human definition of that word, they are among the few birds who can recognize themselves in a mirror. When they find one of their own dead, they hold a "funeral": they gather together calling raucously and then leave all at once in silence. They also recognize individual humans; biologists who've climbed trees to measure magpie eggs have found that the parents recognized them personally on subsequent days and mobbed them, while ignoring other people nearby.

Their natural inquisitiveness has led to their being maligned, to their detriment. It's long been believed that they have a habit of stealing shiny objects and anything else that catches their fancy. But research has shown that, in fact, magpies are wary of unknown objects. It also was believed that they harm cattle and waterfowl, so in the early 1900s there was a bounty on their heads in the western US, and they were slaughtered as pests. In fact, they land on cattle to eat ticks and other insects—helping, not harming, their kin. They are also seen following wolves and coyotes to scavenge from their kills. Opportunistic omnivores like their corvid cousins, the jays, ravens, and crows, magpies mainly eat insects, seeds, vegetation, and carrion.

They're also, if not threatened, quite tolerant of humans. Like Steller's and Canada jays, they frequent campgrounds, suburban areas, and parks. In places like the parking lot at Exit Glacier in Kenai Fjords National Park, Black-billed Magpies are spotted pecking at car bumpers, picking off insects caught on the drive there. For most of us, they are intelligent, beautiful companions throughout Alaska's long, dark winters.

TOM SEXTON

Magpie at Twilight

I watch a magpie walking back and forth
in a puddle that will be frozen by morning.
It pauses to adjust its long iridescent tail.
Three ravens are watching from a cable wire.
They seem to know when it's best to keep quiet.
The magpie turns its head for a better look
at me, beakless, shivering in my down jacket.
"Noah sent me out with the dove," it chatters
before it turns away, falls silent for a moment.
The ravens tilt their heads, chortle in disbelief.

Little Brown Bat

Myotis lucifugus

The Little Brown Bat is the lone bat throughout much of Alaska, living in the southeastern rainforests, the spruce and birch forests of the Interior, and treeless communities in the western part of the state. The state's southern regions have six other bat species in smaller numbers. The Little Browns have round ears—hence their Latin name, which translates as "mouse-eared light-shunning." Able to live more than thirty years, this tiny bat is also the most common and widespread bat species throughout North America, although populations are currently plummeting.

Bats are unique in all the animal kingdom as the only warm-blooded mammal who flies. They're also known for using echolocation rather than sight to feed at night. They're adept at eating insects and can eat hundreds of mosquitoes per hour. One study in Interior Alaska found that, surprisingly, almost half of the diet of Little Brown Bats consisted of orb-weaving spiders. Biologists concluded that in Alaska, as opposed to more southern regions, the bats get more food value from the larger spider bodies than from mosquitoes and other insects.

Lactating females, with high energy demands, can consume more than their own weight (about that of two nickels) in spiders and insects in one night of foraging.

During the day, Little Brown Bats rest. Mother Bats cluster together for warmth in maternity roosts, in Alaska favoring buildings, especially warm attics and furnace rooms. A single naked baby (known as a pup) is born in early summer and is able to fly in about a month. Males and nonreproductive females roost singly or in small groups, not just in buildings but also in live and dead trees, stumps and woodpiles, caves and rock crevices, and under peeling bark. Little Brown Bats, 3 or 4 inches long, can slip into cracks just ⅜ inch wide. When roosting in cold weather, they can enter a state of torpor to reduce their metabolism, heart rate, and body temperature—thereby saving energy.

Very little is known about Alaska's Little Brown Bats in winter—whether they stay or hibernate, and, if they stay, how they survive winters. They have been found in both April and September, suggesting overwintering, and some captured Little Browns have ragged ears, indicating frostbite. Biologists are using both active and acoustic monitoring, as well as attaching radio tags, to learn more.

Little Brown Bats and other bat species elsewhere in the country are in very serious decline for various reasons, including habitat loss, large-scale wind turbines, and the use of pesticides, but especially from white-nose syndrome. Caused by a fungus, white-nose syndrome first appeared in New York State and has been steadily moving west and north, but it has not yet been identified in Alaska. Little Brown Bats, along with eleven other North American bat species, are listed as endangered under the Endangered Species Act.

Humans can help Little Brown Bats, and all bats, by reducing pesticide use and welcoming these insect-eating night fliers with houses attached to poles or the outside walls of buildings. It can be a thrill to see these tiny mammals darting and swooping in the dim light of a summer night, chasing down a meal on the wing.

MAR KA

Little Brown Bats

I've found downed bats in our driveway, little brown bats
called, officially, *little brown bats.* Juveniles, still learning to fly.
As the mosquitoes whose numbers they help to keep down
are now out biting, they'll come out of hibernation soon,
en masse, but we generally won't see them, zipping high
in the small-hour forest shadows, click-click navigating,
roosting singly, in cracks, tree or stone, except for the times
nursing young demand a colony's close-hanging warmth. There
they begin to learn that flying involves falling to launch.

I've found little brown bats in our driveway, made efforts
to keep away the dog, the cat. Can't help with the owls, hawks,
coyotes that go out roaming at night looking for a snack.
The bats must do all right, though, as they live two, three, decades,
twice as long as the predators, six times as long as a mouse,
as a small songbird. Sum, then, greater than the parts. Not dumb—
the world's only flying mammal it chirp eco-locates to chart
its darting way. Order *Chiroptera*, Greek for "hand-wing," a thing
made of skin-cape stretched over four long fingers, a thumb.

I've found little brown bats in our driveway, which reminds me
that as a species, bats are the second most common mammal group
after rodents (think rats). About a fifth of the world's animals
are bats. These, here, abide in the forests into which our homes,
with our kids and dogs and cats, have encroached. No surprise
if I have bat guano on my boots, have tracked it all over the globe.
No surprise if some other animal, wild or not, breathes
in, with the guano-powder, a virus able to, through it, transition
to humans, which we then spread willy-nilly on our airplanes.

The little brown bats I find in my driveway—might our car
have hit them, swooping in, fast, unexpected, at low light?
It's that flying ability, unique among mammals, that sparks
an immune system evolved to repair cell damage, carry,
without effect, layered virus loads. I'm told a hundred million
coronavirus microbes fit on the head of a pin. How many
move daily through a little bat's system, exposing all the pigeons
on the shared windowsills of abandoned buildings? And these
birds leave droppings, to dry and flake on car windshields . . .

Thinking of the little brown bats in our driveway, I make
linked hand-wings that the sun casts in large shadow
on a facing wall. Mystical mythical apparitions materialized
by angle of light. This is not just-somewhere-else, it's here
in our country, in our world, twirling, mostly oblivious,
towards its own truth, its own doom. Maybe a bit of hope
if you're reading this poem to its end now, thinking about
the pursuit of shadows, the shadow's pursuit, about falling
to launch, about techniques for mapping in the dark.

Cat-tail Moss

Isothecium myosuroides

Cat-tail Moss, one of hundreds of species of mosses found in Alaska, is probably the most common moss in coastal rainforests, especially in Southeast Alaska. Cat-tail Mosses are shape shifters: they hang from branches like cats' tails or curtains, grow into mats that look like forests of miniature trees, and drape over logs and boulders. In some forests, they cover nearly every tree trunk and branch. What on first glance can appear to be multiple types of mosses in a forest is very often just multiple forms of Cat-tail. Other common names are Tree Moss, Icicle Moss, and Variable Moss. Their diversity extends to coloring, which can range from whitish green to brownish green and glossy to matte.

As a bryophyte, Cat-tail Moss uses rootlike structures to attach to surfaces; water and nutrients are absorbed from the air, rain, and surrounding water through the leaves. Mosses in general are adapted to survive in many environments; many can go dormant during dry conditions, turning brown and looking dead but coming back to life quickly with water. All mosses must have water to grow, as well as to reproduce, because their sperm need to swim through water

to reach the egg for fertilization. Mosses are also harmed by trampling, invasive plants, human development, and changes in winds and temperature.

Although mosses, including Cat-tail, are full of nutrients, their high content of cellulose and wood-like compounds, some of which are toxic, make them indigestible by most birds, mammals, and insects. One exception is the northern bog lemming, who lives in Southeast Alaska. Another is the microscopic invertebrate water bear, or tardigrade, who lives almost exclusively on mosses; the water bear's talent is using sharp claws to pierce individual moss cells to suck out the contents and leave the indigestible cell walls behind. As well, black and brown bears reportedly eat moss before hibernating, to stop up their digestive systems.

Birds and small mammals use Cat-tail and other mosses for nest linings. Humans use mosses for such things as chinking between cabin walls, insulation, padding, and treatment of wounds—since some mosses contain antibacterial compounds. But these uses are secondary to mosses' tremendous importance to their forest communities, where they provide habitat and support for so many other forest beings, like insects, worms, slugs, fungi, lichens, ferns, and young trees. Mosses absorb and retain rainwater, feeding moisture and absorbed nutrients over time to other plants and preventing runoff and erosion. They enrich the soil as each generation passes to the next. They also play an important role in carbon sequestration, helping to buffer climate heating.

In *Gathering Moss: A Natural and Cultural History of Mosses*, Robin Wall Kimmerer wrote, "Mosses are so little known by the general public that only a few have been given common names. Most are known solely by their scientific Latin names, a fact which discourages most people from attempting to identify them. But I like the scientific names, because they are as beautiful and intricate as the plants they name. Indulge yourself in the words, rhythmic and musical, rolling off your tongue." Say it—*Isothecium myosuroides*.

MISTEE ST. CLAIR

Becoming Moss

For so many years I have lived in this rainforest
watching the rain become more rain.
Cat-tail moss hanging from branches like a carpet of coral.
Thrown over logs like blankets. If green could be gold.

Whenever I found myself in the desert
or anywhere with a strong sun,
I would think *I could be happy here.*
Because I wanted to persist, I told myself
I could grow anywhere but salt.
Because so many rainy days I thirsted
for the sun to plump me. I dreamed sun.
I dreamed sun.

But always there was a thin thread that held me in place.
I thought it meant loosely attached. I thought it meant not rooted.
But lately I've learned to keep low to my own boundary layer.
Something about growing into a place. How slowly I saw
my own simplicity. I never desired an ambitious flowering—
I don't compete well. Now, when I come home from far away,
the air tastes wet on my body. I open my palms to funnel the rain.
Permeable to home. Loyal to this land. It means
I don't transplant well, I know this now.

Pushki

Heracleum lanatum

Pushki, from the Russian for "little cannon," also commonly known as Cow Parsnip, is a tall perennial who can reach 10 feet high, with thick hollow stems and large maplelike leaves that can be up to 16 inches across. The Latin name of this plant couples a reference to Hercules, for the large size, and *lanatus*, meaning "woolly" or "downy," for the stems and leaves, which are covered with hairs. In Tlingit, the plant is called Yaana.eit, "thing that is situating there," perhaps suggesting this being's large presence.

These distinctive plants grow throughout most of Alaska in moist fields, woodlands, and alpine meadows, as well as along roadsides. They're sometimes solitary but more often group in large numbers, creating a beautiful sight. The small white flowers are arranged in large umbels (flat-topped clusters) at the top

of each stem. In fall, the green seeds—which when crushed give off a distinctive sweet-tart scent—turn brown with winter's onset.

Dozens of bird species—including woodpeckers, chickadees, and juncos—feed on these seeds, as do bears, deer, and moose. The latter animals also favor eating the tender early spring shoots. In winter, the dried stalks often remain towering above the snow, frost lacing their umbels like a prism.

Pushki is an important traditional food for Alaska's Indigenous Peoples and was also embraced by homesteaders. The stems, first peeled to rid them of their hairs, can be eaten raw or cooked like celery. The roots have been used medicinally by Dena'ina and were burned as incense to ward off illness. Unangax̂ and Alutiiq treat colds and sore throats with Pushki teas and place heated leaves on minor wounds and sore muscles.

The juices of this being are phototoxic; that is, chemicals in both the leaves and stems are activated by exposure to sunlight, and on a sunny day spatters of that juice on skin can result in painful burns with large blisters. The chemicals (in a category known as furanocoumarins) have a purpose, though; they're the plant's defense against insects, fungi, bacteria, and some grazing animals.

Each umbel on a Pushki plant can contain as many as nine hundred individual flowers—a giant magnet for pollinators. In high summer, these umbels buzz with flies, bees, butterflies, and moths, all happily enjoying a feast of flower nectars.

ANNE CORAY

Heracleum lanatum (Cow Parsnip)

The flowers seem whiter this summer,
more delicate than I remember:
drops of ethereal blood,
the umbels an onionskin tracing
set down with a fine-tipped pen.

What else might I miss in this life—
how many days have I not seen the sky,
soft rags of clouds shining up the blue,
their shadows tumbling casually
over the mountains, while disillusion

like a dark flame burned
my mind's petty length.
I'm tired of human clamor,
smudge and clutter of the world.
Who wants to go on
governed by the same rude horns,
the demagogues, the rabble?

Let the culture fall to its own cravings;
I'm taking up with things divine:
leaf-filter, sheath and fiber,
stalks so tall they often lean
but determine to grow taller,
that ask for only the rain's thin coins,
the soil's nutrient and a decent light.

Volcano

Alaska's landforms are shaped first by fire, then by ice. From Attu Island and the arc of the Aleutians to Prince of Wales Island in Southeast, Alaska's southern coastline is an active, explosive curve along the Pacific Ocean's Ring of Fire. Here, oceanic tectonic plates slide beneath continental margins, liquifying the mantle below into buoyant magma.

According to the Alutiiq, whose word for Volcano is Puyulek, the first two women came into the world by passing through an erupting Volcano. Today's Alaska has 141 named Volcanoes, of whom 55 have been active since 1760. Many are easily visible mountains, like Mount Edgecumbe near Sitka (in Lingit, L'úx, Blinking Top; or L'úx Shaa, Blinking Mountain). Four active, snow-draped Volcanoes on the Alaska Peninsula can be seen across Cook Inlet from Homer to Anchorage—Augustine, Iliamna, Redoubt, and Spurr.

Every minute of every day exhibits seismic activity in Alaska, and eruptions and earthquakes can occur anytime. When Mount Redoubt erupted in December 1989, ash caused all four engines on a jet carrying 231 passengers to shut down; the plane fell for five minutes before crew could get the engines restarted.

The force of Volcanoes shapes Alaska in many ways—giving the conical shape of some of the most prominent peaks, spreading lava and ash over land and sea. Along the Alaska Peninsula, where volcanic ash has fallen throughout the last century, bays have become shallower, altering habitat and changing who lives there. Dig into soil around the region and you can find the layers of gray ash from eruptions. Hiking a forest trail in Kodiak, you'll see light-colored ash from the 1912 eruption of Novarupta in Katmai National Park, the largest eruption in the world during the twentieth century.

Novarupta created the hauntingly beautiful, multicolored landscapes of the Valley of Ten Thousand Smokes within the park. Generating thirty times the volume of lava and ash created by the Mount St. Helens blast, the eruption destroyed several Alutiiq villages, including Katmai and Savonoski; buried the town of Kodiak in 3 feet of ash; and shrouded Alaska and British Columbia. The acidic ash plume ate bedsheets hanging to dry in Seattle and dropped global temperatures by 1 degree F for a full year.

Alaska's Volcanoes continually remind us of dynamic forces greater than ourselves at work shaping our world. They're also a beautiful and otherworldly sight: towering snow-draped mountains with centers so boiling hot that trails of steam slip from their summits.

ANNE HANLEY

Baked Mountain Suite

FIRST MOVEMENT: UP

After an endless march across the valley
then a final push halfway up
the backside of Baked Mountain,
he fell into base camp—so exhausted
he doubted his own purpose.
What could he, an artist, see with mere eyes
that all the scientists traveling with him
could not reduce to graphs and data?

But the next day was a new day.
Tip-toeing past the sleeping others,
past his own pack, heavy as sin,
this time taking only his third eye,
he stole from base camp and started up.
This time he seemed to float
over the windblown ash never sinking
through its rust red mantle to the snow beneath.
This time, in no time, he found himself
on top of Baked Mountain.

SECOND MOVEMENT: ON

As he stood there gazing, gawking
helpful beings materialized.
"The real secret," they chanted
as they blindfolded him,
spun him around and around.
"The real secret does not require
purity of purpose or intent,
talent, energy, knowledge.
The real secret,"
and here they stopped him, steadied him,
pulled off his blindfold,
"is simply knowing where to stand."

CADENZA: HIS OWN/ INSPIRATION

Looking down on Novarupta,
he could feel the great heart of the world
pulsing blood-hot magma
through all its arteries. He could sense
beneath that giant stopper
the forces of creation and destruction
sleeping like babies in each other's arms.
He watched the way the Magi must have watched
hovering over the manger, awed, abashed,
delighted, hypnotized by the raw potential.
Then he heard himself singing a lullaby,
"Everything that ever was, still is
or is waiting to be born again."
Feeling whole and suddenly hungry,
he headed down.

THIRD MOVEMENT: DOWN

He plodded back down Baked Mountain,
sinking in with every step; ash
spilling over the tops of his boots.
When he got back to base camp,

the others were up, brewing coffee.
They nodded, relieved to see him,
slightly miffed that he, lowest in the pecking order,
was the first one up. He took his eggs, said nothing.
But a young geophysicist with a dusting of milk powder
in his beard said, "What were you doing up there
by yourself?" Having no good alibi, he simply said:
"What else are the tops of mountains for?"
They let it go; a few nodded. Then they all
packed up their instruments and started
off to gather more data.

Boreal Forest

Alaska's boreal or northern forest (also sometimes called taiga from the Russian for "land of little sticks") is often portrayed as one huge expanse of spindly spruce trees covering most of Alaska's interior. While black and white spruce are the most common trees in the ecosystem, the whole is a much more complex tapestry that includes birch and aspen groves, meadows and ponds, willow and alder thickets, and a rich understory of plants including prickly rose, Labrador tea, and highbush cranberry—all of which support beings from the smallest nematodes to the largest moose.

The boreal forest, north of which lies the treeless tundra, is shaped by long, cold, dark winters and short, light-filled summers, as well as by permafrost (permanently frozen ground) and forest fires. Compared to tropical and temperate forests, the boreal is characterized by both a relatively low number of species and a relatively low abundance of individuals. Dramatic seasonal swings in daylight and temperature may well be the most significant aspect of this forest.

Summers are intense, with more than twenty-one hours of daylight at summer solstice and temperatures up to 100 degrees F. They're also brief, lasting only a few months, so both plants and animals grow and reproduce rapidly and prolifically during the long days. Understory plants, comprising most of the boreal forest plant diversity, waste no time in leafing, flowering, and fruiting. Millions of birds from more than 150 species—including alder flycatchers, yellow-rumped warblers, and ospreys—arrive in spring to feed and breed, then return south for less-extreme winters.

Beings that remain year-round make amazing adaptations to bitter winter conditions lasting up to nine months, when daylight drops below three hours on winter solstice and temperatures can plunge to −94 degrees F. Some, like marmots and collared pikas, hibernate. Bears enter a state of torpor that's much like hibernation. Wood frogs and many insects actually freeze their bodies. Birds and mammals that remain active all winter insulate themselves with thick feathers or fur; many, including arctic foxes, lemmings, and ptarmigan, also burrow into snow for even more insulation. Caribou, who summer on the

Arctic tundra, seek forests for winter food and protection. Some small birds, like white-winged crossbills and redpolls, have specialized throat pouches they fill during the day with seeds to digest during long, cold nights.

In summer, lightning-started wildfires have always been a feature of the forest, and plants have various strategies to survive them. They may regrow from unburned rhizomes or from seeds that can blow long distances or remain dormant in soil, awaiting favorable conditions. Fire opens the cones of black spruce, seeding new trees.

The vegetation and soils of the boreal forest store huge amounts of carbon, helping to mitigate climate change; altogether, this forest—stretching from Alaska through northern Canada—is Earth's largest terrestrial carbon storehouse. However, as the climate warms, wildfires have increased in number and intensity, resulting in new patterns of postfire succession, including the conversion of forest to grassland. The climate-warming emissions from fires further worsen the hot, dry conditions that drive the fires. Permafrost beneath the forest is also thawing, releasing carbon from soil that's been frozen since the last ice age.

While unsustainable logging of Canada's boreal forest is a serious threat, timber harvest except for local use has not been significant in Alaska's boreal. The small size and low quality of trees there, along with the distance to commercial markets, have so far limited large-scale logging.

Over a span of thousands of years, Alaska's Indigenous Peoples, primarily Athabaskans, have lived comfortably in the boreal forest. They have developed a deep understanding of the forest's edible and medicinal plants, know the habits and movements of animals, and can use their knowledge of snow and natural materials to stay warm and safe. Today, tens of thousands of humans live within Alaska's boreal forest—in Fairbanks, a city of thirty-two thousand, as well as in numerous villages and on remote homesteads. These residents maintain close ties to what the forest provides.

Consider, once again, winter in the boreal. When snow piles several feet deep and temperatures plummet, this northernmost forest holds a deep quiet. Often, the only sound is the squeak of booted footsteps on dry snow. The heavily needled branches of black spruce bend, holding some snow and dropping some in great piles. On snow-laden boughs and inside the cavities of older trees and snags, beings like red-breasted nuthatches and red squirrels gather against the cold. At the base of trees, snow wells form, providing refuge for winter residents and entries for those who burrow beneath the blanket of white to find warmth.

Boreal Chickadee

Poecile hudsonicus

Most of Alaska's birds ebb and flow with the seasons, appearing in spring and departing in fall. But not Boreal Chickadees. Astonishingly, these tiny brown-capped fluffballs, weighing less than the weight of three pennies, remain in their boreal forest home year-round. As the Gwich'in say, Ch'idzigyek shìk ch'idlìi ihtth'ak—"I always hear chickadees singing."

Their common name in English is an onomatopoeia: it imitates their song. It's also an onomatopoeia in Yup'ik, as the chickadee's voice, Cikepiipiiq, is also the Yup'ik name for the bird. They're known as Uksullaq in Alutiiq.

How do these diminutive singers survive winter, when temperatures can drop below −50 degrees F for weeks at a time? We're not entirely sure, but we do know some of their survival secrets. In summer and fall, they cache insects and seeds in bark crevices and under the rough edges of lichen, and will recall hundreds of these cache sites when the food is needed. Also in fall, they put on heavier plumage, creating their own down coat. They gain fat more quickly than songbirds south of Alaska, up to 8 percent of their body weight each day, and then use this extra fat to keep warm. But winters are so long that these birds, weighing a mere third of an ounce, have to do more than add fat and feathers. During the nights of deepest cold, they gather together in tree cavities and do what's called regulated hypothermia: they conserve insulating fats by lowering their body temperature by almost a quarter, from 108 degrees F to just 85 degrees F.

Boreal Chickadees are among the few songbirds found almost exclusively in boreal forests of Alaska and Canada. But they aren't the only chickadees surviving Alaska's winters with amazing adaptations. Alaska is home to four of the seven species of chickadees: Boreals, black-capped, chestnut-backed, and gray-headed. Black-capped, whose range includes the contiguous US, are the most common and survive winter by being up to 25 percent larger than their southern relatives. Gray-headed chickadees are the rarest; for decades, their populations have been in steep decline, and they are now found only in the Arctic National Wildlife Refuge.

Boreals are rarer and less often seen than black-capped chickadees. But while their populations are smaller, their secretive reputation is partly because—with their rich brown cap, small black bib, blue-gray wings and tail, white cheeks, and cinnamon flanks—they're well camouflaged against the mature spruce, willow, and alder thickets they inhabit. It's also because, unlike most other chickadee

species, they don't vocalize to defend breeding territory, which can be up to 13 acres. Instead, Boreal Chickadees chase intruders, weaving and banking like the most adept of flyers, and threaten with a "ruffle display": they puff up their feathers, make a gargling sound, and jerk their chins up.

Outside of breeding season, Boreals are gregarious. Mating for life, they remain together, joining family groups and other year-round residents like nuthatches, juncos, pine siskins, creepers, downy woodpeckers, and other chickadee species. In spring and fall, migrating songbirds—orange-crowned warblers, ruby-crowned kinglets, yellow warblers, Lincoln's sparrows—join these mixed flocks in search of food and shelter. It's not uncommon for backyard birders to be charmed by mixed flocks of year-round residents and migratories—a colorful mélange of tiny beings filling the air with song and the trees with wings.

Boreal Chickadees are tightly associated with large mature spruce trees for all seasons of their lives. Even their mating displays begin at the top of giant spruce, as the male chases the female in a downward spiral. Logging of mature spruce forests, industrial activity, habitat destruction, and climate-change-driven events like wildfires and spruce bark beetle infestations all threaten their future.

Active, acrobatic, and agile, these tiny, perky birds enliven the forests of Alaska. It's nothing short of a delight to see Boreal Chickadees weaving around spruce limbs and willow branches, hover-feeding and darting among the trees like sprites, keeping us company year-round.

JOY HUNTINGTON

Chickadee, tell me your stories

How do you survive
in snow and ice
while we hide and recoil?
The darkness
invites depression
to stay awhile

Chickadee, tell me your secrets

How do you defy
the natural order of survival?
How do you float and loop
through the frozen sky
unbothered by forty below?

Your small body is determined
to thrive in the death grip
of Alaskan winters

Chickadee, sing me your song

It reminds me that my skin
will not fail
to keep my blood warm

You call out a high-pitched melody
Familiar birch trees respond
They invite you back like family

You endure without drama
You follow the faintest ritual of memory

Seek shelter
Eat for the long winter

Fly
Mate
Nest
Repeat

Chickadee, you do not
ask for permission
to speak

Your song is a celebration of survival
An act of defiance of what it means to be
small and fragile

You make hardship
look joyful

Your tiny wings
soar above
the daily trials
of larger beings

Black Spruce

Picea mariana

The Black Spruce, one of Alaska's three spruces, is the most common tree in Interior Alaska, dominating entire landscapes. (The other spruces are white and Sitka, and a hybrid of the two is a Lutz spruce.) These spindly conifers have little taper from bottom to top and have been called Dr. Seuss trees due to their scraggly, often twisted appearance. The four-sided needles grow singly out of branches and have sharp tips. The roundish, purplish cones are small, about an inch long, and tend to cluster near a tree's top. The bark is notable for its flakiness.

Black Spruces live in very harsh conditions, either in nutrient-poor soils in acidic peatlands (where they're sometimes called Bog Spruce or Swamp Spruce) or in cold, wet soils underlain with permafrost. Their shallow root systems inhabit an active soil layer that can be less than a foot thick. Because of such poor anchoring, Black Spruces tend to blow over in high winds. And as soil freezes and thaws, the trees sometimes tip sideways—leading to the term "drunken forest." These trees grow very slowly; a thirty-year-old tree can be less than 2 inches in diameter.

Black Spruces are typically the first trees to move into the sphagnum mats of drying wetlands. There, by adding organic matter to the peat soil and increasing its fertility, they prepare the soil for other plants. Wherever they live, they provide homes for red squirrels and grouse. Hares use them for cover and feed on their young shoots. Birds nest in them. Caribou also use their forests for cover and feed on the lichens that grow in their shade.

As less benign kin, spruce bark beetles drill through bark to lay their eggs. Beetle kills have increased with a warming climate throughout the Black Spruce's range, affecting mostly mature trees. More than other spruce species, though, the Black Spruce is particularly effective in using pitch to "pitch out" the invaders.

With sap-filled branches, Black Spruces are very susceptible to fire, which usually sweeps through their forests at intervals of between 50 and 150 years. Fires are beneficial to them, as the heat opens their cones and releases their seeds. Without fire, some seeds remain in their cones for up to 20 years.

In the Dena'ina Athabascan language, the word for all three spruce species is Ch'vala, which is also the word for tree and indicates this one's prominence and value. A variant of the word is Ch'vach'etl'a, which translates to "shriveled tree" and refers to the Black Spruce more specifically. The closely spaced tree rings (due to slow growth) make a hard wood the Dena'ina used for many things, including sled runners, spear shafts, digging sticks, tongs for moving hot rocks, and splitting wedges.

Spruce roots from all species were used as twine, for fishing lines and snares, in weaving baskets and nets, and, in general, for tying things together. The Dena'ina and other Indigenous Peoples have used sap for healing burns and sores and as a wellness tonic, and the cambium (inner bark) as a food, eaten

raw or ground into a flour. The pitch had and has many uses, including as glue and caulk and applied to wounds to stop bleeding.

Black Spruces grow close together, in such communion that when one is killed by spruce bark beetles or cut by humans, the others are much more susceptible to not only beetles but also windfall. This degree of connection is apparent during a mast year, when trees produce a heavier-than-normal crop of cones and seeds. The mast years of the Black Spruce and their neighboring white or Lutz spruce synchronize cone production over a surprisingly wide swath of forest, such as the entire Kenai Peninsula. The wealth of seeds overwhelms the ability of seed predators, increasing the chances of survival for each seed.

This wax-and-wane cycle is why red squirrels, who depend on spruce for more than half their diet, practice hoarding; each squirrel can stash up to nine thousand cones a year. But in mast years, even such intense hoarding can't keep up, and many seeds survive to grow new trees.

DEBBIE CLARKE MODEROW

Winter Spruce

In darkness, final flakes swirl. A gust delivers one last wail. Then, quiet. The blizzard surrenders, the night sky responds. One star, then two introduce a new story. Moon, nearly full, illuminates the boreal forest, changed.

Deep snow settles in waves. Buries tracks of red squirrels near the woodpile, patterns of wings and talons etched in blood-stained snow. Drifts blanket the forest floor—its tundra, willows, dwarf birch. Hundreds of black spruce preside over a shroud of white. Slender trunks rise from deep wells, snow halos sculpted by yesterday's wind.

These elder spruce—seeded a century ago when wildfire swept the valley—pose, weathered and scraggly. Shallow roots cling to frozen ground. Their branches laden with snow, some trees tilt. Others bow. Elegant in white, nimble and strong. As if their needles had not browned. As if the spruce bark beetle had not thrived here. As if each tree were not dead. Or dying.

Spruce shadows bend, moon moves along. Elongated ghosts shift and fade. A boreal owl sounds a few notes. Curls of woodsmoke weave through spruce. Standing, still.

Canada Lynx

Lynx canadensis

The Canada Lynx is Alaska's only native wild cat. Rarely seen, called ghost cats for their secretive behavior, these beings are nevertheless ubiquitous throughout Alaska's forests. Their name comes from the Indo-European word for light, *leuk,* referring to the luminescence of their big, reflective eyes. The Dena'ina of Southcentral Alaska call this being Kazhna, meaning "black-tailed one," in reference to the black-tipped tail. In Dena'ina tradition, Lynx is a shape-shifter, able to also be human, who teaches not only prowess as a hunter but also responsible and socially acceptable behavior.

One of four species of lynx worldwide, Alaska's Lynx are, like all cats, curious, playful, and intelligent. Similar in size to their cousins the bobcats, who roam areas to the south, Canada Lynx are distinctively made for the Far North's snowy winters: they have longer ear tufts, longer legs, and longer, thicker coats. But what most sets them apart is their paws—as big as a cougar's, although cougars are four times larger and heavier than an average eighteen- to thirty-pound Lynx. The extreme size of a Canada Lynx's thickly furred paws allows them to travel over deep snow almost as easily as their main prey, snowshoe hares, and to support twice as much weight on snow as bobcat paws allow.

Their big paws also make them excellent climbers (not unusual for a cat) and swimmers (very unusual for a cat, whether wild or tamed). No known natural barrier, save the oceans, can stop their journeys. They cross Alaska's tallest divides, from the Wrangell Mountains to the Brooks Range, and even Alaska's widest, strongest rivers, including the Yukon, Copper, and Kuskokwim. While these beings, who can live to fifteen years old in the wild, are often content to travel about a dozen miles a day in search of food, they also range so widely that one we see in Alaska may also live in Canada.

Some of their forays are epic, rivaling the great migrations of caribou. One adult male (known as a tom) collared by biologists at Tetlin National Wildlife Refuge traveled in the course of a year through the Yukon and Northwest Territories of Canada, eventually settling west of Great Slave Lake, 2,100 miles away. Another tom, collared near Bettles, traveled 550 miles across the Brooks Range to the Chukchi Sea near Icy Cape, beachcombed for 200 miles, doubled back along the 500-mile route to the Dalton Highway, then meandered the Brooks Range to finish off the year at the headwaters of the Killik River.

What makes them roam? Most likely, it's the decadal population cycles of the snowshoe hare. The relationship between Canada Lynx and snowshoe hare populations is a classic example of predator-prey cycles: when snowshoe hares are abundant, Lynx populations grow; when snowshoe hares decline, so do Lynx, with a lag time of about two years. This cycle occurs in sync across their range, so that as the peak of the hare cycle travels in a wave across Alaska and Canada, Lynx follow this wave of abundance for hundreds of miles.

Lynx also alter their litter size in sync with snowshoe hare populations. When hares are abundant, females, known as mollies, can successfully raise as many as six to eight kittens. But when hares are scarce, mollies have fewer kittens or forgo having litters at all.

Canada Lynx are in decline over most of their range; not only are they losing habitat through logging, industrial activities, and human development, they're also facing illegal hunting and liberal commercial trapping laws in Alaska. As well, Lynx are so tightly conjoined to snowshoe hare populations that as climate change threatens the hares, it also threatens the Lynx.

Obligate carnivores, Lynx must eat meat to survive. So when snowshoe hare populations dip, they turn their candescent eyes to grouse, ptarmigan, squirrels, voles, lemmings, and muskrats. They also gravitate to backyard chicken runs, which even when covered in netting might not keep those chickens safe. Ghost cats, on their big, quiet paws, will find the slightest opening.

CHRISTINE BYL

Lynx

Lynx. Listen: *lynx*. A feline, a cat. A verb (this links that). A word without vowels, and an *x* to boot. Lynx whose population spikes so closely mirror snowshoe hares' that you can't think *cat* without thinking *rabbit*. Lynx travels fastest at night, hunts on the ground but can climb trees and swim. On my road, a few miles down from the cabin. In the park, asleep midday, skulking the ditch line after dark. *Lynx*. The *l* rolls out shy, almost hidden in the tongue's crevice, the crick of the middle *nk*, as if poised for a leap, and *x*'s secretive hiss, falling on tufted ears. Such bony structure, visible beneath the skin of animals and words.

Wandering Fleabane

Erigeron peregrinus

Wandering Fleabane, a member of the aster family, lives in the Pacific Northwest, from Alaska to Oregon, along stream banks and in moist mountain meadows. Each plant puts up a single erect stem, usually with one flower at the top, that can grow to 2 feet tall or more. The narrow leaves are longest near the base and get shorter as they climb the stem. The flowers of many thin rays are purple or pink but can also be blue or white, with yellow centers. This fleabane is also sometimes called Coastal Fleabane, Subalpine Daisy, or Wandering Daisy (although it is not a daisy).

Why "fleabane"? Why "wandering"? "Fleabane" comes from Old English and refers to the plant's odor, which supposedly could repel fleas. Historically, fleabane was burned or dried in sachets to repel annoying insects, although there's no scientific evidence supporting use as a repellent. The species name, *peregrinus*, is Latin for "traveler" or "wanderer" and presumably relates to this

plant's spread. The genus name, *Erigeron*, translates to "old man in the spring" and refers to the hairy stem.

Fleabane's nectar provides a feast for a diversity of butterflies, flies, and bees, who are apparently not deterred by the smell. Other insects or their larvae feed on the leaves or flowers, and spiders set up their webs on Wandering Fleabane's scaffolding. When the yellow centers go to seed, they become food for seed-eating birds, who also feed on the small beings that feed on Fleabane. Various Indigenous Peoples have used Wandering Fleabane medicinally—for example, the Alutiiq have treated colds, chest congestion, and pneumonia with it. The plant's leaves have antioxidant properties and can be eaten either raw or cooked, like spinach.

Wandering Fleabane is one of 390 known species of fleabane in the world, of whom 170 live in North America. Several close kin live in Alaska, including the one-flowered fleabane, bitter fleabane, fringed fleabane, arctic fleabane, purple fleabane, Alaska fleabane, and Denali fleabane. Other plants are sometimes called fleabane although they don't belong to the *Erigeron* genus. The beach fleabane, one of these, is also known as seaside ragwort and is a member of the *Senecio* genus. One lesson we can learn from fleabane is the importance of learning the Latin names of plants, to separate often confusing and overlapping common names from the scientific specifics!

MARTHA AMORE

Wandering

Here you are again,
petals crowding like
too many crooked teeth,
your lavender tarnished by
a center of filthy yellow,
daisy-like but somehow
not at all like a daisy.
Unlovely fleabane,
your name suits you well
among the wildflowers.
Fragile red lanterns of
columbine and
dainty lady's slippers,
bending stalks of chiming bells
and bushes lush with
cinquefoil and Sitka roses and
wild geraniums,
all bunch and gather like
so many cliques of beauties
along high school halls.
Yet here you stand,
fleabane,
spindly in your
mismatched colors,
splay-toothed and
bold and
grinning.

Gray Wolf

Canis lupus

Through deep winter snow, they move in single file, leaving behind a wavering line across the whiteness, the alphas in front shouldering the way to ease the passage of youngers behind them. Gray Wolves inhabit almost all of Alaska's terrestrial habitats, navigating harsh Arctic winters and wet Southeast summers with equal resilience.

Wolves are among the most social of all nonhuman vertebrates, with incredibly strong family ties. They live in family groups, also called packs, led by an alpha pair whose monogamous bond is at the heart of wolf social organization. Gray Wolves—who weigh between 50 and 150 pounds and who are about 2½ feet tall—go to tremendous effort to remain with their families; relocated adults have traveled hundreds of miles to return. They also try to help other family members caught in traps or snares, and they return to sites of trapping deaths for weeks.

Wolf social organization is rooted in cooperation, both in rearing pups and hunting prey. Family groups can be long-lived: the Toklat family group in Denali National Park was at least seventy years old when it was decimated by trapping outside park boundaries. When allowed this longevity over generations, each Wolf family group develops unique learned behaviors, from hunting to pup rearing to communicating, that are finely tuned to their territories.

Gray Wolves, whose coats range in color from pure black to gray-brown to all white, are quasi-nomadic in winter, following moose, deer, sheep, and caribou herds, and den centered in summer, often using the same den year after year. Dens may be used for hundreds of years: Denali's Toklat River den that was used when Adolf Murie studied them in the 1940s was also used as recently as 2012 before the alpha female was killed by a sport hunter.

A Gray Wolf den is an elaborate, honeycombed series of deep burrows and entrances with a maze of interconnecting trails, all spread over many acres. Dens may originally be excavated by ground squirrels and later enlarged by foxes and then Wolves. When Wolves aren't using them, dens may host foxes, ground squirrels, porcupines, and wolverines. Archaeological evidence indicates that at least three ancient Wolf den sites in Denali were also shared with humans from three thousand to ten thousand years ago. Humans co-evolved with Wolves in a reciprocal relationship.

Raising new pups at a den is a family affair, with nonbreeders cooperatively nursing, pup sitting, teaching, and guarding pups. In territories with more challenging prey like Dall sheep, pups require a two-to-three-year period—about a quarter of their life span—to learn from more experienced Wolves. Yearlings pup sit while the rest of the group goes hunting; older Wolves take pups on short walks to acquaint them with the world beyond their natal den.

Wolves play with each other many times a day and howl frequently, expressing a range of emotions and communicating, often to locate one another, announce their return from a hunt, indicate when they or other members of the group are in pain or distress, energize the group after a rest, and advertise territorial boundaries. Their howls enliven the cold stillness of the Arctic night.

Gray Wolves are deliberate about which animals to pursue, and also depend upon scavenging. When together, they hunt larger prey, including Dall sheep, moose, and caribou. But they also hunt smaller mammals and even birds: in peak snowshoe hare years, Gray Wolves shift to them. Wolves enhance the health of prey populations by targeting the weakest.

Scavenging of winter-killed ungulates can contribute up to 85 percent of Gray Wolves' diet in harsh winters. They commonly dig as deep as 10 feet into hard-packed snowdrifts or even avalanches to reach frozen carcasses. They then eat parts as they thaw, returning to the site over time.

Ravens lead Wolves to prey and follow them to scavenge the remains of their kills. Many smaller animals like ermine also scavenge Wolf-kill remains. Wolves compete with brown bears, stealing these beings' kills and harassing them to keep them away from moose calving areas.

Alaska is home to between seven thousand and eleven thousand Gray Wolves, and in most of the state they are not, as their Lower 48 relatives are, considered endangered, at least not on a population level. However, family groups adjacent to human habitation are extremely vulnerable. Primary threats are habitat loss from clear-cutting, human development, hunting, trapping, and predator control campaigns. The latter three factors have a greater negative impact on Gray Wolf families than natural mortality and rarely improve ungulate populations. Whereas natural losses usually affect young, inexperienced Wolves, shooting and trapping take more alpha adults. Loss of alphas can cause loss of unique hunting abilities, constricted territory, decreased socialization, loss of pups, and ultimately fragmentation of the entire family group.

With their long-lived family groups, Gray Wolves provide continuity and balance to Alaska's wild places. And they can be a delight to witness: on a sunny summer day, high on a promontory above rolling forest and tundra, a young Gray Wolf lies on her side while several fluffy pups clamber over her, delighting in their wild world and awaiting the return from the hunt of the rest of their family.

NICK JANS

The Circle of the Kill

I looked up, and a wolf stood in the ice fog. Tail out, stiff legged, he regarded me, and our eyes met. Something in the stare of a wolf is chilling, beyond the yellow eyes of childhood nightmares, beyond any threat, real or imagined. Caught in that cold glare, I felt suddenly transparent, as if my heart were being measured. In that instant I knew what we fear most in wolves: not their teeth, but their wisdom, an alien, elusive intelligence that refuses us, rejects our notions of superiority with a glance.

The wolf turned. I followed his gaze and there were dark shapes in the brush, a dozen wolves staring down toward us, waiting for what would happen next.

Then he was gone. In a blur I felt more than saw, the wolf pivoted and launched himself uphill. Ahead, the pack was in full flight, heads out, tails back, merging into single file as they burst up the slope. On the skyline they took a last look and vanished into the land.

Halfway up the slope I found the kill. Entrails, bones, clumps of hair, and bloody snow were all that remained of the moose. The story of the hunt was there in the snow. The snow was trampled in circular patterns over several hundred yards, and at each place, the cow had made a stand. The intertwined trails seemed graceful, as if the wolves and moose had danced together. At the end of their dance lay the kill, beautiful in its simplicity. Here on this silent hillside, there was no horror. This was their life—an endless hunt, an endless celebration of death.

As I stood within the circle of the kill, looking down at wolf prints frozen in blood, I brushed against their secret: wolves understand death perfectly. That's the bright, cold wisdom we see in their eyes, the thing that makes us afraid. Death is their art, their beauty, while it's our darkness and terror. If we ever understood what they know, we've forgotten. Maybe we're drawn to them because they remember.

Aurora Borealis

The Aurora Borealis or Northern Lights, that display of light in dark northern skies, is one of the delights of Alaska's winters. Cold nights find Alaskans and visitors alike out on darkened porches or remote hilltops far from artificial lights to watch the show of undulating ribbons and curtains, mostly in shades of green but also in swirls of yellows, reds, and purples. This atmospheric phenomenon has been called the holy grail of sky watching.

The Aurora is created between 50 and 300 miles above Earth's surface when energized particles emanating from the sun collide with the upper atmosphere at speeds of up to 45 million miles per hour. Earth's magnetic field serves as a protective barrier and redirects the sun's charged particles (called the solar wind) toward the poles. (In the southern hemisphere, the phenomenon is called the Aurora Australis or Southern Lights.) The energy transfer from the particles to atoms and molecules in the atmosphere excite them to higher energy states. When they relax to lower energy states, they release their energy in the form of light, resulting in colors that depend on specific molecules in the atmosphere; green is produced by oxygen molecules, red by oxygen at higher altitudes, and purple by nitrogen. The shapes and motions of the Aurora are caused by the constantly changing input from the sun, responses from the upper atmosphere, and Earth's own motions.

Study of the Aurora tells scientists about the density, composition, and strength of electrical currents in the upper atmosphere, which together tell them about the Earth's magnetic field and space weather, which can sometimes disrupt technologies.

Auroras vary over time, depending on the sun's emissions, which go through cycles of activity. When the sun's storms of energy are greatest, Auroras are most frequent and brightest. The sun's highest rate of activity, known as the solar maximum, most recently occurred in 2024. Auroras take place year-round, day and night, and their activity is forecasted by a number of agencies and institutions, including the University of Alaska Fairbanks Geophysical Institute.

Auroras have fascinated people forever, with most northern cultures connecting them to ancestors. In a 2016 film made by scientists and Iñupiat elders, *Kiuguyat: The Northern Lights*, elders shared that the Aurora was a welcome source of light; when it was bright and strong, the ancestors were said to be happy. Other times the Aurora was seen as threatening, as a force that could come down and harm people. The Tlingit have believed that the spirits of those who died during warfare were active with the Aurora and that displays were signs of impending war and bloodshed. In Alutiiq cosmology, the Aurora (Qiugyat) has also been believed to be animated by the spirits of dead warriors. Both the Alutiiq and the Yup'ik have thought that whistling brings the lights closer.

Do Auroras make sounds? For years scientists disputed this, but recent studies have confirmed just how this occurs. With a temperature inversion on a cold night, electrical charges can build up in the inversion layer (within a few hundred feet of the ground). When an Aurora increases in intensity, the geomagnetic disturbance can reach down and spark an electrical discharge in the inversion layer that can be heard as faint rustlings or clapping.

Whether through sound or sight, Auroras remind us that life on Earth wouldn't be possible without the shield of the planet's magnetic field, protecting the surface from harmful cosmic radiation from the sun. Most of all, beyond their scientific fascination, the shimmering, undulating curtains of light we call Auroras are among the most astonishingly beautiful celestial events to be witnessed across Alaska's cold, dark winter skies.

TRIPP J CROUSE

Bagone'giizhig (Hole in the Sky)

Berry brambles line
a road paved
in grief,
up the Milky Way;

An afterlife sojourn,
    one day, I hope
    to return.
And if so
    I hope to see
    an aurora phoenix
    guiding me home.

Electric kisses
    hum between pinpoints,
like a song
    slipped from the brackets,
crystalline daggers
    with ragged edges,
shattered fragments
        of light
    slip and slide
    on waves
    of sapphire magnetic
            solar knives.

Ermine

Mustela erminea

In summer, the sleek, sinuous Ermines are reddish brown with white chests and bellies. But come winter, they turn entirely white except for the black tips of their tails. Like ptarmigans, arctic foxes, and snowshoe hares, Ermines change color seasonally to blend into their environments and hide from predators. The transformation is triggered by variations in day length (the photoperiod), which influence the levels of melanin in new fur as it replaces what's shed. The black tail tip is another evolutionary adaptation, thought to cause a predator—owl, hawk, fox, lynx—to reach for it and thus miss the rest.

Ermines are a type of weasel, sometimes called a Short-tailed Weasel or, in Europe, a Stoat. They live throughout Alaska except on some islands and are a close relative of the smaller least weasel, also found in Alaska. They have short legs and triangular heads. Adults are just over a foot long, with males larger than females. Their lives are short—one or two years—but females reach sexual maturity quickly, within three or four months. They birth an average of six or eight "kittens" in dens, where they may also dig storage areas for excess foods.

Being so lithe, Ermines can slither through tunnels and burrows under snow to catch their primary prey, small rodents like voles and shrews. But since they must eat 40 percent of their body weight every day year-round, they also go after birds, fish, and even insects, as well as scavenging from the kills of wolves and bears.

For their size, Ermines are surprisingly fierce—even challenging people who separate them from food. Their bites, relative to body size, have more force than those of bears. The combination of size and ferocity has given Ermines an outsized role in myths, ancient and modern, around the world. In Alaska's Indigenous cultures they're often associated with wealth or good luck, and their white winter coats or tails are used to adorn parkas and other clothing.

Overhunting and trapping, as well as habitat loss from development and clear-cut logging, have reduced Ermine populations. Shifts in climate are also cause for concern as other competing species move into Ermine habitats. What's more, as winter's snow arrives later and is less consistent, Ermines' coat-change to white makes them more visible and susceptible to predation.

It's surprising, in a vast snowscape, to see a small head pop up from a burrowed snow tunnel, only the black eyes and nose distinguishing the Ermine from winter white. As the sine wave scampers away, the only marker of this being's passage is the black-tipped tail, bobbing to disappearance.

NANCY DESCHU

Circle of Seasons

Three weasels race across the trail
through summer's narrow tunnel.
Running black-nose-to-tail,
nose-to-tail, then the last black tail.

Long and sleek, brown back, white belly,
so lithe, a leaping line, a wheel in motion.
Mere weasels in summer, judged as ruthless killers,
(a single bite to the back of a chicken's neck!)
circling to lustrous white ermine in winter,
killed for the robes of royalty.

One, two, three, so fast they pass,
perhaps just innocent play
as they dive into the alder forest
where hares and voles freeze in place,
breathless.

Chaga

Inonotus obliquus

Take a walk in a forest of Alaska birch and you're likely to spot Chaga, a parasitic fungus. The hard, irregularly shaped body is dark and roughly textured, resembling a patch of burned charcoal stuck to the side of a tree. This body is not the fungus's fruit but simply a mass of mycelium, a rootlike structure that absorbs

nutrients from the environment to feed the fungus. When the mass is broken off, the inside is colored orange or gold.

Chaga takes its name from the Russian, which in turn comes from the Indigenous Khanti language of Siberia. Chaga has many other names—among them Clinker Polypore, Cinder Conk, Black Mass, and Birch Canker Polypore. (Most polypores are woody shelf fungi with pores instead of gills on their undersides.) In England and Canada, Chaga's official name is a long one: Sterile Conk Trunk Rot of Birch. Chaga also has nicknames—"black gold," "mushroom of immortality," "diamond of the forest."

Chaga is not benign, but the damage is slow: the spores enter a tree through wounds and cause decay within the host for up to eighty years before killing the tree. While the birch is alive, only those dark sterile masses—the visible Chaga—are produced. However, the tree's growth is stunted and often deformed. When the infection kills the tree, fruiting bodies grow beneath the bark; these are rarely visible but can spread the infection to other trees.

The National Institutes of Health have called Chaga "a super-fungus with countless facets and untapped potential," although understanding of its mycochemical (natural antioxidant) composition and biological activity is limited. Scientific findings have demonstrated Chaga's abilities as an antioxidant, anti-inflammatory, antiviral, and antitumor agent. Chaga is currently one of the most intensely researched medicinal fungi and is especially prominent in cancer research. Northern Indigenous Peoples have used Chaga medicinally for hundreds, if not thousands, of years to boost immunity, fight infection, and for other health purposes. To make tea, they typically break the Chaga mass up into small bits or grind it into a powder, add this to water, and boil, then simmer it.

Chaga harvesting and use have caught on throughout Alaska in recent years. At least four thousand commercial products utilizing Chaga (including tinctures and supplements as well as teas) are currently available, nearly all of them sourced from wild harvests. The impact of overharvesting may be of concern but has not been studied.

Aside from human uses, Chaga, like all decomposers, benefits other beings and overall forest diversity, particularly through nutrient regeneration, soil aggregation, and water retention. Chaga is also a food source for animals large and small.

Chaga is only one of Alaska's many kinds of fungi: about fifteen hundred mushroom-producing species have been named, but scientists suspect that more than five thousand different species exist. While Chaga, like Alaska's other species of shelf fungi, grows on trees, many more mushroom species are found growing on the forest floor. Among these are the beautiful but deadly fly agaric or amanita, the edible shaggy manes boletes, and chanterelles, and the russulas with their many colors.

DAWNELL SMITH

Love Medicine

Chaga and Birch wed
among boreal coverts
during a feast of insistence
burnt steadily into scar

their embrace porous
like a poultice, like passage
way, like kin medicine weeping
the heart's food from winter's wake

a sacrament held
within decayed tree rings, split
carapace, the ancient knowing
yoked to its rekindling fate

Boreal Owl

Aegolius funereus

The Boreal Owl is silent except from mid-February to April, when the repeated *who* call of up to twelve notes resounds through the northern forests. One of ten owl species found in Alaska, Boreals live only in northern forests and nest in tree cavities (or nest boxes provided by humans).

With a dramatic facial disk, white eyebrows, and brilliant yellow eyes, this small brown-and-white bird—only 10 inches tall—can look either surprised or annoyed. But to see a Boreal is rare. In daylight, Boreals roost high in trees, hidden in canopies and camouflaged against dark trunks. At night they hunt, but, unlike other owls who swoop around, Boreals drop straight down quickly on prey from perches.

A Boreal pair is usually monogamous and has the most extreme size differential of any owl, with the female much larger and heavier than the 3.8-ounce male—sometimes twice as heavy. In April or May, the female lays between three and seven eggs. She does the incubating, and the male brings her food for herself and the chicks. Boreals prey largely on voles, but also shrews, mice, and sometimes small birds and even insects.

Owls have asymmetrical ears—one lower than the other—to help them acoustically pinpoint the location of prey by judging height as well as distance. The asymmetry of the Boreal's ear openings is greater than that of any other owl species. This owl can hear a vole even under snow and will plunge through snow layers to capture it. As a year-round resident of the north, a Boreal often caches food in crevices or the forks of trees and then thaws a meal by sitting on it.

One Yup'ik name for this owl is Takvialnguaraq, which translates to "one with poor eyesight," probably referring to the bird's daytime sight. Other Yup'ik names refer to its call. The *Aegolius* of Boreal's scientific name is Greek for "bird of ill omen," and the Latin *funereus* relates to "funereal," with its implication of death. In numerous mythologies, owls are often associated with bad luck or death—perhaps because of their mystery and sometimes-sudden appearance at night, or their wailing calls. In general, Alaska's Indigenous Peoples have associated owls with the supernatural, sometimes as bearers of news, both good and bad, and sometimes as powerful spirits.

Like other owls, Boreals regurgitate the bones and fur of their prey as pellets, usually once a day. A good way to find a nest is to look for woodpecker holes in dead trees and then at the tree's base for piles of small pellets. Or follow their call, which sounds much like the winnowing of a snipe's tail feathers—a low, repeated *who-who-who-who-who.*

NICOLE STELLON O'DONNELL

Lonely Owl

Evenings
early and late
under the star-black sky
punctuated
by satellites
the boreal owl calls
from somewhere
in the aspen.
Five beats.
Not *who*,
but *whowhowhowhowho*,
meaning *Why am I still alone?*

His feathered loneliness
was so much like yours
when you went out

to feed the chickens
and look at the stars,
and looked back to see me
through the living room window:
laptop open,
face lit with green light,
so unaware of you.

You stood on the deck
hooting at me: five beats
copying the owl in his sadness
calling me outside to see
the aurora and stars
and breathe in the cold moment
with you.

I would love to say
I heard you,
but I didn't.

I never even looked up.

Alaska Paper Birch

Betula neoalaskana

The trunks of Alaska Paper Birch—usually white with multicolored patterns of gray, yellow, and red—tell stories. With peeling and curled bark, they change texture and color with the passage of time. They harbor dark claw marks of black bears who've climbed them, and they host whole ecosystems of lichens, mosses, and fungi.

Alaska is home to multiple and hybridizing *Betula* species, but the most common is *neoalaskana*, very similar to *Betula papyrifera* (the paper birch most common in New England and the Great Lakes Region). The difference between the two is on the chromosomal level and is not obvious to an observer. Most Alaskans simply call this being Paper Birch or White Birch. "Paper" describes the bark, which can tear off in sheets.

This Birch grows most densely in Alaska's interior and usually in mixed forests with white or black spruce (who follow it in a forest's succession after fire or clearing). The trees can reach more than 60 feet tall and have more than one trunk. The oval leaves are toothed and pointed. A single tree bears both male and female flowers (catkins), long and short; the females form fruits with tiny winged seeds that often litter the snow in winter. Those same seeds are a major source of food for redpolls and other flocking winter birds. Roots are shallow, generally less than 2 feet deep, and associated with mycorrhizal fungi aiding tree nutrition and soil health.

Because the bark is both flexible and waterproof and can be made into baskets, dishes, and coverings of all kinds, this being has always been highly valued by Alaska's Indigenous Peoples. Another name for this tree is Canoe Birch, for the use of the bark as canoe covers. A person should be very careful in harvesting bark, though, as even peeling a bit of loose bark for a fire starter can leave the tree vulnerable to attacks by insects. Girdling a tree (removing bark all the way around it) will kill it. The best source of bark for human use is a downed tree or pruned limb. Aside from the bark, a Paper Birch's wood is an important source of material for structures, tools, and firewood.

Alaska's Indigenous Peoples have also traditionally used Birch leaves to make a comforting tea and to treat illnesses. Young leaves and catkins can be added to a salad. The outer bark contains chemicals that show antiviral, antimicrobial, and anticarcinogenic properties; these are currently in pharmacological testing. The inner bark (cambium) served as a "famine food" in late winter for many Indigenous Peoples. In spring, Birch sap, like that of maple trees that grow in more southern regions, transports nutrients from roots to branches where new leaves will form. Historically, the sap was collected by Indigenous Peoples to drink as a health tonic; made of sugars and minerals, it was one of the first carbohydrates available in spring. The sap can also be boiled into a syrup.

The warming climate is creating problems for Alaska Paper Birch, primarily with drought and insect infestations. In recent years, two invasive birch leaf miner species have caused Birch tree leaves to brown in late summer, sometimes killing entire trees outright.

As one of Alaska's few deciduous trees, Alaska Paper Birch is notable for fall's color changes. These trees, along with cottonwood and willow, provide a spectacular golden color against the dark green of spruce, lighting up the forest even as the days grow shorter.

ERIN COUGHLIN HOLLOWELL

Choir hive

Against white sky, the birch tree opens
its many dark
                    mouths. She hears its words,
golden river under snow. Secret
honey. Clapper of vein-scribed marble,
that bell rings
                    each full moon.
                                        Now waning.
Now feeding the mountain underneath.

Northern Red-backed Vole

Myodes rutilus

Northern Red-backed Voles, often mistaken for mice or shrews, are small rodents with fuzzy coats, gray on the belly side and reddish or rusty on the back. They have short tails and are about 5 inches long. Each weighs about the same as a deck of cards.

As slight as they are, all voles play an outsized role in the food web as the primary prey for a diversity of other beings including foxes, coyotes, martens, weasels, owls, hawks, other large birds, and even bears, wolves, and northern pike. Put simply, these unassuming beings convert plant nutrition and energy into carnivore nutrition and energy.

Millions of voles, in ten species, live throughout Alaska. The Northern Red-backs are among Alaska's most ubiquitous and common animals. They are found everywhere in the state except south of the Stikine River and on some islands. Red-backs also inhabit northern Canada, northern Russia, and Scandinavia.

Red-backs typically live in or around forests and prefer dense ground cover for protection from weather and predation, but they also live in tundra and among rocks on overgrown talus slopes. They create surface runways through vegetation as travel corridors, and they can climb trees much like squirrels. They are most active at night and so are not often seen by humans. They remain active year-round and live in family groups that share grass nests and intricate networks of tunnels under winter snow and layers of insulating moss. When given the opportunity, Red-backed Voles also happily overwinter in warm human buildings.

The omnivorous Red-back's diet consists of berries, grasses, leaves, shoots, buds, roots, fungi, lichens, and seeds, supplemented with insects and carrion. Like other voles, they gather and store food in their nests for winter consumption. They help disperse plants and fungi through excretion, by caching seeds, and by relocating plant parts. They also improve soil quality by aerating soil, and they encourage the growth of beneficial mycorrhizal fungi.

These beings are terrifically prolific. One female can have up to four litters of six to eight young every year, and young can begin reproducing when only four weeks old. On the downside, an individual rarely lives more than a year. Red-backs go through population booms every four or five years, providing more food for their predators and thus increasing those populations as well.

Harsh winters, especially those with extreme low temperatures and little snow cover, cause their populations to crash.

Like all rodents, voles have highly specialized long whiskers called vibrissae that are vital tactile organs for all aspects of a vole's life—including finding food, sensing wind direction, measuring distance, and communicating with each other. With vibrissae and scent marking and occasional squeaks, Red-backed Voles live their short lives tightly woven into their families and home places.

ERICA WATSON

Relentless

The vole stood wet and shaking on the plastic lid I'd left baited with stale trail mix and floating in a five-gallon bucket, holding a peanut in its tiny clawed hands.

I'd drowned half a dozen garden-eating voles this way already that spring and felt no moral qualms until now. "The only thing I'd feel weird about," I'd said earlier that morning, "is finding one still alive in the bucket," and with those words I'd manifested this conundrum. The vole's black beady eyes looked up at me, asking directly: *what now?* In the garden bed above, a row of radish sprouts stripped of their leaves by tiny teeth grew barely above the soil, and seed-sized poops marked a trail down the wooden frame to the bucket resting at its edge.

Resentful, I scooped the vole into a dry bucket and drove it across the river, where it wouldn't find its way back and would feed an owl or a lynx. One year, a neighbor had marked the voles he'd live-trapped in his house with nail polish, and the same ones kept returning; a geographical barrier was needed. "I never said I was a pacifist!" I insisted as I drove away, and it was true; this action was more cowardice than conviction. I had not spared it from death, only delayed it and denied the role of my own hand, my own bucket.

"The only true defense is deterrence," I said, nailing scrap wood and hardware cloth across gaps in the greenhouse, realizing too late I'd borrowed the language from the US Border Patrol. It's embarrassing how a three-ounce rodent can make anyone sound like a cop. How their relentless and insatiable pursuit of food reduces their human neighbors to our most squeamish and petty, their eyes a mirror of our failed attempts to build an impermeable world.

Bohemian Waxwing

Bombycilla garrulus

The Bohemian Waxwing is one of Alaska's most brightly colored birds. In winter, flying in large flocks, they're a rainbow of color in a monochrome landscape. These midsize crested birds with black masks, red and yellow wing spots, and yellow-tipped tails look like they're edged with crayon markings or wax—hence the common name "waxwing." The colors come from the carotenoid pigments in the fruit the birds eat. As the birds get older, the waxy bits grow larger.

Bombycilla comes from the Greek and Latin words meaning "silk tail." *Garrulus* is Latin for "talkative." Waxwings, unlike most songbirds, don't have a true song but communicate with high-pitched trills. If they're in a tree near you, you will hear them! The "Bohemian" in the common name is thought to relate to the fact that the flocks wander widely, as is said of the nomadic Romani people of Bohemia. The related, slightly smaller and duller-colored, cedar waxwing (associated with eating the berries of cedar trees) rarely ventures as far north as Alaska, but a stray occasionally joins one of the Bohemian flocks.

Bohemians nest in conifer forests in Interior Alaska, where they can often be spotted perched on the top of a black spruce, ready to intercept flying insects. Both males and females gather nesting material and feed the nestlings, but only females build the cup-shaped nests in which they lay and incubate two to six eggs. The mating ceremony is very elaborate:

the male and female pass food back and forth to one another again and again, up to fourteen times.

In winter Bohemians congregate in the southern and southeastern parts of the state, where they roam around in large, noisy flocks to find and feast on berry-bearing plants and trees. Native species with berries that hang on into winter include mountain ash, juniper, and highbush cranberry. The birds have an uncanny ability to find fruit and will descend on trees almost anywhere to quickly strip the berries within hours and move on. In the past fifty or so years, the planting of ornamental trees like the European mountain ash in Anchorage and other Alaska cities has become a strong lure to the Bohemians, who flock into neighborhood fruit trees in noisy hundreds.

While they're able to metabolize the alcohol in fermented fruit, the birds also can become intoxicated—and behave drunkenly. Fruit is rich in sugar but lacks nutrients, so a single Bohemian will eat hundreds of berries in a day, amounting to twice their own weight. Because flocks travel widely and each bird defecates every few minutes, they distribute fruit seeds over great distances.

The Bohemian is considered a common bird, although one in decline. North American populations have fallen by 55 percent since 1970. While ornamental plants and trees provide significant food in Alaska and elsewhere, the draw into cities makes the birds vulnerable to window and vehicle collisions (especially after feasting on fermented berries). Pesticides used on plants and trees are an additional threat. And where invasives like starlings continue to expand, Bohemians are losing ground.

Not much is known of these gypsy birds. While in winter they're frequently seen and heard, their hidden summertime lives in the forest and the great fluctuations of their numbers from year to year make them a mystery.

SUSANNA J. MISHLER

Throw Salt

over your shoulder so the dead stay
near us. So they never float
skyward in their last clothes and disappear

into the atmosphere. Now
crystals wink from the floor
and the table, recalling the taste

of blood and earth. The birch
outside is rooted in earth and air—
a vascular connection

between kingdoms.
Which is why trees often look like ghosts.
Why, when a flock of waxwings

erupts from the mountain ash,
they take the yard with them.
Not coordinated in their turning,

like shorebirds, but strung like a loose
signature over morning's blue page, parts
drawing away and back to an unseen

line, other parts arcing to an altogether
different edge. It's hard
not to think of this spectacle as love

from beyond, a miracle
of fertilizer. Of how one day my thumb
could be a feather, a beak, a berry.

Wood Bison

Bison bison athabascae

Wood Bison are the largest terrestrial animals in North America, with males standing 6 feet tall at the shoulder, stretching 10 feet long, and weighing as much as 2,000 pounds. Yet a century ago, these impressive creatures—the larger of the two kinds of modern bison found in North America, the other being the more southerly plains bison—nearly vanished from the face of Earth. Also called

Mountain Bison, Wood Bison thrived in Interior Alaska for thousands of years until they disappeared in the early 1900s, likely from unregulated hunting and a changing environment, about the same time as musk oxen vanished.

The massive bone structure and musculature in Wood Bison's large forward hump allows them to swing their heads deep into winter snow to forage for vegetation beneath. Their dark brown hair is not hollow (as is the hair of moose, sheep, and caribou), but more like human hair. They begin to shed their coat in spring and by summer have completely replaced the old coat with new.

Females mature at two years old and give birth to a single reddish-tan calf in spring. Newborn calves can stand within minutes of birth and can run and kick within hours. Although they continue to nurse for months, they begin grazing within a week or so. Adults graze in meadows, around lakes and rivers, on sedges, forbs, grasses, and willows. They form herds of twenty to sixty individuals and move seasonally to optimal foraging habitats.

Wood Bison were once thought to be entirely extinct, with the last known native Wood Bison in Alaska shot in 1915, but a small number were discovered in a remote area of Canada in the 1950s. Conservation efforts there have since increased their numbers to more than ten thousand. Reintroducing Wood Bison to Alaska became a government goal, with the expectation that returning them to portions of their former range can restore their ecological role: by grazing, wallowing, and fertilizing, they encourage plant growth and diversity and create habitat for other beings.

In 2008, the first group of these wild beings from a herd in Alberta, Canada, was trucked to Alaska. After being held in captivity until 2015, 130 were flown and barged to the lower Innoko/Yukon River area of western Alaska. These herds are not, however, doing as well as humans had hoped, primarily because ice layers within heavy snowpack—winter conditions caused by climate change—keep the Bison from being able to reach forage. In one winter alone, 70 percent of the Wood Bison died. In 2024, additional Wood Bison were introduced to the Minto Flats State Game Refuge west of Fairbanks.

Plains bison, never native to Alaska, were introduced to the state in 1928 near Delta Junction, thousands of miles from their natural range. Four herds now number about nine hundred individuals, who are subjected to permit hunts. Some plains bison, better than cattle at avoiding brown bears, are also held on ranches, including on Kodiak Island.

In only a few places now in North America, these massive and majestic beings can be heard bellowing to each other, a call that's somewhere between a purr and a growl, deep and resounding.

DARYL FARMER

Wood Bison Trampled to Death by Own Herd

A runt already broken and bruised even before
being transferred with 39 others to the Large Animal
Research Station. What prompted the stampede?

A car door slam? The howl of wolves? So many sounds
to spark fear after two days huddled in a dark trailer,
and then released disoriented to a new temporary pasture.

I'm told it was a dog barking from the nearby trail,
and think of all the mundanities that can inadvertently kill.
Once, an elderly man driving down the middle of a gravel

road forced me to the road's edge and my car skidded
into the ditch. In my rear view, I watched him drive
slowly away, oblivious. What damage we do in our

plastic-filled days, our pesticide nights, our diamond-
mine dreams. I decide to name the bison Claudius.
These wood bison, once extinct in Alaska for over

a hundred years, have been transported from Canada,
a restoration that will transform the entire landscape,
feces attracting ants attracting birds dropping seeds

germinating green sustenance and on and on, effects rippling,
a project to admire. Still, I am haunted by the harsh act,
though I know it is the fault of no one.   Claudius,

tonight I mourn for you, and I mourn the dog who barked
but meant no harm and I mourn that last Alaska bison death
years ago, and I mourn for all of us who have ever driven

straight ahead without looking back. And I mourn
the ants at that ditch's edge whose hill
was destroyed by the tire they never saw coming.

Fireweed

Epilobium angustifolium

Midsummers in Alaska blaze with the brilliant pinks and magentas of Fireweed, a tall, showy wildflower that fills entire meadows with color. Fireweed especially thrives in lands that have been recently transformed by natural forces like wildfires and flooding but also grows along roadsides, streamsides, and forest edges throughout most of Alaska.

This being's common name in fact comes from their ability to quickly colonize landscapes burned by fire. Fireweed does this in two ways. First, their fast-growing rhizomes usually survive fires, sending up new shoots that can bloom within a month of a canopy-clearing fire. Second, those new shoots produce an explosion of seeds that easily blow into areas that may lack rhizomes. By stabilizing soils and cycling nutrients, Fireweed prepares landscapes for other plants to follow in succession.

Fireweed's stalk, which flowers from bottom to top, works like a calendar for Alaskans, marking summer's progression. The bottommost buds open in midsummer, and the topmost blossoms open as fall approaches. As the leaves turn pleasing shades of orange, red, and purple, the flower pods close up into sheaths. When the sheathes pop open with cooler weather, the white silken seed strands float off with the wind. Alaskans say, "When Fireweed turns to cotton, summer's soon forgotten."

A very close relative, *Epilobium latifolium*, known as dwarf fireweed or river beauty, prefers gravelly places along rivers and on rocky mountainsides around arctic ground squirrel burrows, on disturbed and fertilized ground. This being grows much closer to the ground, with powdery gray leaves and larger flowers.

Bees—wild and domestic—love Fireweed, and beekeepers on the Kenai Peninsula consider Fireweed the major source of their bees' nectar. You'll find Fireweed honey, jelly, and syrup at farmers markets and local shops throughout Alaska. Bees help Fireweed by cross-pollinating; they habitually start at a bottom flower and work upward, carrying pollen from a top blossom on one plant to fertilize the bottom flower on another.

Many other beings get nutrition from Fireweed. Bears, famished after a long winter, browse the early shoots, and deer feed on stalks. Moose, caribou, muskrats, and hares all get a taste.

Fireweed enjoys a long relationship with humans, with both traditional and modern uses. The tender spring shoots are as delicious as young asparagus and packed with vitamins A and C. Iñupiat sometimes mix them with seal (or other) oil to preserve them for winter use. Young leaves and buds are mixed with other greens in salads or vegetable dishes, and tea is made from steeped leaves. Fibers from Fireweed stems have been wound into twine, and "fluff" from seedpods has been mixed with mountain goat fur in weavings.

Like many northern perennials who face brief summers, Fireweed emerges as red shoots early in spring. The red color helps to absorb sunlight and protects against frost damage. From these first shoots to their flowering to their fall leaves, these iconic plants, often referred to as Alaska's unofficial state flower, brighten the landscape with color.

CHAUN BALLARD

A Poem Ending with a Strambotto Wherein I Include an Extra Line That Is Myself *or* A Poem in Which I Name the Flower

I think back to the words of the geo-spatial engineer,
        sitting next to my wife & myself in a bar, who shared
with a curious neighbor about the renewing nature
        of fire. How, when it comes to fire, it is healthy
for the land to set itself against itself. For context,
        I am not speaking of country. Currently, I am speaking

of fire. & speaking of fire, my buddy returns home from
the years he tallied upstate. &, even I, if I may venture
a tad off course, to enter into the weeds, am tempted
to call his return to "civility" *freedom.* But the true
freedom is not in my naming him *returned,* but how
my friend emerges from a front door that he unlatches
himself, to record what grows freely in the absence
of one man. & like that, it is dandelion season again.

A resident weed of every state. & my favorite
poet reminds me that a weed cannot be contained,
even in the hubris of this poem. So I admit, I am
often ashamed of how I can pass between several
towns without the practiced ritual of honoring one
field with a stolen image. My wife points out the fire-
weed. &, like a child, I repeat *Fireweed,* without taking
notice. Here, from where I am currently perched,
are entire fields purpled with fire. & my buddy, who
has emerged from the doorsteps into the yard, is speak-
ing into the camera's beating heart to posit how *You
can eat this weed* referring to a species of vegetation unlike
that which delivered him upstate. & tangential, perhaps,
is how one species of plant might devour another.

& how one may facilitate another's growth. You see,
the thing about fire is that my friend is now vegan.
A raw foodist to be exact. & my wife's uncle Jerry,
who used to bring her dandelion jam throughout
the years of her youth, purchased his first suit
to attend our wedding. I'm talking, years ago—
on an island where we watched the slow hands of lava
seduce a flame that coursed down a hill's green back,
until all that was left was absence—which, yes,
is a precursor to growth. &, upon this growth, I would
have sworn to you that every wild thing that bloomed
was a flower. Which is absurdity, I know. My hubris
flaunting its purple head. The stuff of bees collecting
in the calyx of fireweed like the curls of one's hair.

More than likely, Uncle Jerry is buried inside the fabric
of his first suit—with the petals of something beautifully
wilted in his lapel. & listen to how I desire to say *fire*
when there is no need. Speaking of summer, when
the allure of redwoods sheltered us from the burn
with their leaves, we walked between laughter & astonish-
ment: Uncle Jerry, & his identifying every species of plant
by their scientific root. Now, my own father is dying—
while I wish for a summer when I feared the bees
who were too busy to fear me back. Today, I watch them
land & exit this poem. Because living is hard work.
Especially, when every piece of legislation is meant
to uproot you. &, at the heart of every grass root is
a movement of bees fighting to keep. Which is a statement.

& believe me, I understand the effort it takes to show.
& not explain. So, as a gesture of new growth, allow
me this one moment to re-catalogue *a tall showy wildflower*
that grows from sea level to the subalpine zone. A member
of the Evening Primrose family, taxonomists previously included in
the willowherb genus. Today, *it is now* home in *the Chamerion*
group. Which is code, yes, for *fireweed*. A wildflower known
for its verse-like persistence. & by persistence I mean:
It is appropriate to begin *with a song.*

Crowberry

Empetrum nigrum

Among Alaska's abundance of berry-producing plants, Crowberries are unassuming, almost hidden in the bogs, heaths, tundra, and alpine slopes where they thrive, not only in Alaska but around the circumpolar north as well. Depending on their location in Alaska, they are also known as Blackberries and Moss

Berries. In Dena'ina, the name is both Dghilingek'a ("mountain berry") and Gigazhna ("berry black").

The berries are not exactly black, but they're darker and duller than the blueberries they often grow among, and their leaves look like prickly mosses. The plants, who belong to the heath family, are adorned with small pink or purple flowers in spring. They're often found in rocky soil—hence the *Empetrum* ("on rock") part of their botanical name. The low patches of plants catch soil blown by wind, and their rootedness prevents erosion, making them good stabilizers of mountain slopes.

Crowberry plants guard their places. When the leaves fall off and begin to decompose, they not only build soil around the plant but also release toxins into the soil to prevent the growth of other, competing plants. It may be those chemical properties that give Crowberry medicinal uses. Leaves, stems, and roots have traditionally been made into tea; in the Kodiak region, smoke from burning the plants was believed to cleanse homes and visitors of disease and evil spirits, and the smoke is still used to rid houses of insects.

Crowberries are an important food for many beings with whom they share home places, including bears, songbirds, waterfowl, pikas, voles, and insects; they are especially valuable for grouse and ptarmigan. It's unclear whether the common name comes from their being eaten by crows or a belief by some that this berry is good only for crows—not humans—to eat.

Indeed, Crowberries are seedy and acidic so aren't favorites of human berry pickers. But they've been an important survival food and source of vitamins, fiber, and antioxidants for northern and especially Arctic people. Usually picked after a frost, as cold sweetens them, these berries are made into sauces and jellies or baked into cakes and muffins, often mixed with blueberries as an extender. They're a common ingredient in akutaq (sometimes called Eskimo ice cream) when mixed with fat or oil and bits of dried fish or meat. An Iñupiaq dish, tingaulik, blends them with fish livers.

Crowberries are a constant, holding tight to their treeless home grounds. Their small, hard berries cling to evergreen stems throughout winter and are still there in spring thaw, persistent and humble.

NICOLE STELLON O'DONNELL

Picking Crowberries

Late this fall we pick crowberries
because the blueberries are gone,
skins split from frost, wrinkled purple.

You pull branches up, hoping,
but we have come too late.
The crowberries hang low on moss,
still firm. We settle for them,

too seedy for anything but syrup.
I kneel by the plank path and reach
for piney stalks. I am the new arrival,
dropped in the woods with you.

My fingers feel newly attached, spindly, clean.
Yours feel certain. As I drop the first handful,
they bounce across the bottom of the bucket.

At the truck stop, the waitress
sees our stained hands, smiles, "Berries?"
"Yes," we say, forks poised above pie,
ready to break the crust.

Milbert's Tortoiseshell

Aglais milberti

The beautiful and wide-ranging Milbert's Tortoiseshell Butterfly, with a 2-inch wingspan, is among the most commonly seen of Alaska's more than eighty species of butterflies. Jacques-Gérard Milbert, for whom this butterfly was named, was a French naturalist who collected specimens in the United States in the early 1800s. "Tortoiseshell" is most often applied to cats, referring to the coloration of tortoise or turtle shells, but in this case describes a butterfly's wings. Another name for these butterflies is Fire-Rim Tortoiseshell, which perfectly fits their gorgeous appearance: the upper side of the wings is black-brown dotted with red-orange, with a wide outer band of yellow-to-orange rimmed by black. When their wings close, blue spots on the hind wings become visible.

Milberts are especially noticeable in spring, flitting everywhere through the state south of the Arctic (as well as throughout most of the contiguous United States). Milberts regularly alight on the ground, rocks, or trees, where they rest with open wings—a practice thought to help them absorb sun and keep warm. They typically live near water or wet areas and are often observed in the open, along trails and roadsides. Their populations vary widely from year to year, depending on weather conditions.

Females lay large batches of eggs—as many as nine hundred—on the undersides of leaves and may have two broods in a season. The caterpillars, just over an inch long, are black with white speckles and spikey tufts; the spines mimic thorns and help deter predators. In their youngest stages, the caterpillars group up into webs; as they molt into larger individuals, they spread out more to feed on their host plants—typically nettle but also willow and birch. Eventually they find places to hide (often inside a folded leaf) and pupate; their chrysalises are a beautiful brown-green with a coppery sheen.

As adult butterflies, they feed mainly on the nectar of flowers but also on sap and rotting fruit. They remain active into the fall, then go dormant for the winter under tree bark as well as in woodpiles and sheds; they also overwinter in the chrysalis form. In spring, overwintered adults reappear with the sun, often looking a bit tattered.

Males are territorial and will chase one another. A Milbert's poised on a log or rock below a hilltop is likely a male waiting for a female to flutter along. Another clue: males are generally smaller than females and are more brightly colored. If you get close enough to see an abdomen, a male's includes a pair of clasping organs used in mating. Males also tend to emerge before females in spring.

Milbert's is one of only six butterfly species in Alaska who can produce antifreeze compounds to prevent ice crystals from forming in their tissues during freezing temperatures, when they go into dormancy. No wonder seeing this bright sunrise-colored being in early spring seems like a miracle.

JOHN MORGAN

November Surprise

Fairbanks, Alaska

Ten below and ice-mist on the river
when "Oh," she says, "a butterfly!" as it
comes wobbling from the sun-room, settles
on the floor. We offer sugar water
in a spoon and watch its sucking tube unroll.
It sips, then flutters to the windowsill
and folds its scalloped wings against the chill.

By noon, bright sun, and full of spunk it beats
against the glass, in love with light. The ground
outside, a spanking white, looks welcoming.

Its wings, like paisley, red and brown, quiver
as it paws the pane, embodiment of
summer in late fall, cold-blooded thing,
whose hopes will never be this young again.

Snowshoe Hare

Lepus americanus

In winter's deep snow, the Snowshoe Hare is completely camouflaged. Only the slightest movement reveals this small mammal's presence. Then, in a flash of quick hops, the mirage is gone. The Hare has escaped on the large hind feet that support this being like snowshoes. Webbed and thickly furred, these hind feet allow the Hare to hop across even the most powdery snow without sinking in. Snowshoe Hares are among the most agile and athletic of hares; they evade predators by running up to 30 miles per hour as they dart this way and that, changing direction so quickly that their pursuers often lose them in the chase.

One of two hare species in Alaska, Snowshoe Hares (also known as Varying Hares) thrive in the boreal and coastal forests of mainland Alaska. Arctic hares, who also go through seasonal color changes but are three times the size of Snowshoe Hares, inhabit the western coast and tundra north of the Brooks Range. In Iñupiaq, Snowshoe Hares are known as Ukalliq; in Gwich'in they're Geh, and in Koyukon they're Gguh.

In the fall, like "termination dust" on the mountains (the first dusting of snow, foretelling summer's end), the Snowshoe Hare's feet and ear tips tinge white, and over a couple of months this snow color spreads until the whole body is nearly invisible. In spring, the transformation is reversed, and they become brown furred again, blending into the mixed spruce, wooded wetlands, and shrubby areas where they live.

An important link in the food chain, Snowshoe Hares are the major food source for lynx, as well as for fox, coyote, marten, owls, and other birds of prey; they've also been a traditional source of food and clothing for Indigenous Peoples. The populations of their predators, especially lynx, fluctuate with populations of Snowshoe Hares, who follow a roughly ten-year cycle of boom and bust.

Unlike rabbits, who don't exist in the wild in Alaska, Hares are precocial: born ready to survive on their own. Called leverets, baby Hares are born with their eyes open and a full body of fur, and can hop around the nest in the time it takes their fur to dry. Within weeks, they're weaned and eating on their own.

Snowshoe Hares eat leaves and grasses in summer, and bark and stems in winter. They also visit mineral licks daily and often eat soil; this geophagy (earth eating) may help maintain mineral balances and neutralize plant chemicals that inhibit digestion. Winter snow lets them reach higher for food, finding tender new growth on alder, birch, and willow.

Defending territories of up to 10 acres, Snowshoe Hares follow scent-marked pathways year-round. Their sense of smell is so keen that they can maintain their trail network even after a thick snowfall. Against winter's cold, they pull their long legs under them, lay their ears back, and fluff out their fur. And they rely on snow's insulation, snuggling against snowdrifts and into snow cavities around tree trunks.

But climate change is causing problems for Snowshoe Hares: in winter, with less snow and more rain, they stand out in their white coats and become easy targets for predators. Climate change-induced wildfires are destroying their forests. They're also losing habitat through human development, mining, and oil and gas exploration.

Snowshoe Hares are crepuscular—active at twilight—and nocturnal, but in Alaska they can't depend on the cloak of darkness during summer's long days. Camouflage and silent movements hide them, but they make noise when they want: they growl to defend their home range and thrum their huge hind feet, much like the springtime wing drumming of ruffed grouse, the sound reverberating through their sheltering home forests.

ERIC HEYNE

Death and the Earth's Weather

Pelt mottled gray and white, April's
snowshoe hare is exposed both on
and off the melting snow.

Castaneda's
and Casanova's sly creations—
those two Don Juans, the shaman
and the ladies' man—took their cues
from death, the big one and
the little: One must have both
a powerful will and a stiff world
on which to impose it.

We love to conflate climate
and weather, like the global
warming that made it rain last November.
But green-up wasn't late
this year after all,
just right
on
schedule. Heat waves back at you.

This is the same train we rode as children
up in the dome car, folding our beds down
in the sleeper on our wedding night,
listening for the crossing whistle
up late with the baby. Same train,
same snowshoe hares, same death
in the stern wake of a wavering jet stream.

Wood Frog

Lithobates sylvaticus (formerly *Rana sylvatica*)

North of Yakutat, Alaska has only one cold-blooded amphibian: the Wood Frog. There are no reptiles in Alaska (except an occasional garter snake who may have traveled down a transboundary river from Canada) and only six species of amphibians (two salamanders, two frogs, one toad, and one newt), all of whom cluster along the warmer southern coast of Southeast Alaska. All, that is, except the tiny Wood Frog, who grows to be just under 3 inches long and lives only a few years. These frogs have found a way to survive Alaska's subzero winters so well that they survive north of the Arctic Circle.

In the fall, when nights dip below freezing, Wood Frogs nestle into leaf litter near wetlands, creating small divots called hibernacula where they remain, motionless, until spring thaw. For a few weeks they freeze at night and thaw during the day, and during this period they convert glycogen in their livers into glucose. When winter's freeze sets in, two-thirds of their body water turns to ice. Their hearts stop beating, their blood stops flowing, and their glucose levels soar. These high levels of glucose keep water inside their cells and the frogs alive.

In spring, because they're so close to the surface, Wood Frogs thaw quickly. Returning to normal functioning within twenty-four hours, they find the nearest body of water—typically ephemeral wetlands (also called vernal pools)—to begin their short breeding season. Within a week of thawing, Wood Frogs finish laying eggs. This quick transition allows tadpoles, who eat algae and decaying plants, to transform to adults, who use long, sticky tongues to catch insects, spiders, worms, snails, and slugs, all before the vernal ponds have dried up.

While this freeze-thaw cycle is their most astonishing adaptation, they're also unique among frogs for their dependence on woodlands, their range of habitats, and their long-range movements. By far the most widely distributed frogs in Alaska, Wood Frogs are also widely distributed in North America, ranging as far south as the peatlands of North Carolina. In Alaska, their core range is the boreal forest.

According to the Koyukon, long ago a Frog-Woman was treated badly and killed by two boys. Afterward another boy found her body and buried her. From then on, he was rewarded by good luck in hunting. The little marks on the Wood Frog's back are said to be scars remaining from Frog-Woman's mistreatment. Ever since, the Koyukon have considered the Wood Frog to be good luck, with strong healing powers. When someone has a headache, Koyukon elders say to put a Frog atop their head; the Frog's throat's up-and-down movement will take pain away.

Because of their dependence on a contiguous variety of habitats, especially ephemeral wetlands, and because these wetlands are imperiled by habitat loss and climate change, the future of the Wood Frog is far from certain. But for now, in early spring, vernal pools throughout Alaska still ring with the looping quack-like calls of the males as they sing another season to life.

MARYBETH HOLLEMAN

skating after many moons

it's a small lake, a pond, really, or not
even that, a wet meadow in summer
where our one kind of frog who contains
some kind of antifreeze in their veins lives,
the only one can survive this far north,
our only amphibian, only cold-blooded being,
and when he was young, my boy, and I wanted
to keep him awake to the world, we came here
and listened to the frog chorus, recorded it
for a scientist cataloging the city's frogs
who wanted to see how changing weather
was changing them, would there be more
or less, would they die out as summers dried.
she gave us a recording so we'd know
what to listen for, but there was no mistaking
the tender high notes, held and overlapping
like the rounds I sang at summer camp, calls
unlike the looping lilt of birds. now when
I stand on this frozen meadow, in white
figure skates I got when I was just a girl still
in easy love with this world, I hear all that's held
under ice, since we recorded the frog chorus,
my boy grown to a man who may or may not
notice the frogs are leaving, when I stand here
seeking the certainty a child feels about
the world that holds her steady, a mother feels

for her boy holding her hand as they cross
the damp meadow and stand still, listening,
when I move, clumsily, cautiously, afraid of
reeling headfirst onto what might crack, my
limbs begin to remember, from my girlhood,
the few times the lake froze, and my dad
took all us kids there, swung us around, and
that one time I skated with a friend who flew
circles and arcs around my leaden feet, my limbs
fling it all and swing me around and around
until I am dizzy in love again, until the world
is aright and all this keeps me twirling.

Rivers, Lakes & Wetlands

Look at a map of Alaska and you'll see it veined with blue lines of major rivers and spidery threads of streams, and marked with curved blue ovals of freshwater lakes and patches of small grass bunches that indicate wetlands. Although Alaska makes up only 17 percent of the total area of the United States, it accounts for one-third of the country's surface freshwater. The state's freshwater ecosystems spread from the rainforests of Southeast to the vast Interior forest and the Arctic coast, and from high alpine lakes down to sea level.

Alaska has more than three million lakes larger than 5 acres in size, more than twelve thousand rivers, and hundreds of thousands of streams and creeks. Wetlands cover 43 percent of Alaska's huge terrestrial area. More than a hundred thousand glaciers and 4,000 square miles of icefields (constituting one of the largest nonpolar icefield complexes on Earth), along with lakes, rivers, groundwater, streams, and springs, connect Alaska's uplands with marine waters.

Alaska's largest rivers are the Yukon, Kuskokwim, Susitna, and Copper. The powerful Yukon, flowing for more than 2,000 miles from Canada and then through Alaska, is the third longest river in North America. Alaska's rivers support many beings, including both resident and anadromous fish, and provide passage to the networks of smaller waterways for fish spawning, rearing, and overwintering. The same rivers and tributaries provide streamside habitat for terrestrial beings and transportation systems for humans.

Lake Iliamna is Alaska's largest lake and the second largest freshwater lake that lies entirely within the United States (after Lake Michigan); it is 80 miles long and 25 miles wide and drains eventually into Bristol Bay. It's home to a small population of freshwater harbor seals—one of only two such populations in the world (the other being at Lake Baikal in Siberia). Lake Iliamna also supports many other beings, including the largest numbers of sockeye salmon of any lake in the world.

Glacier melt feeds into many streams and lakes, carrying vast amounts of pulverized rock commonly called rock flour. These sediments, suspended in the water, are nutrient rich and turn lakes and some rivers beautiful turquoise colors when sunlight hits the particles. In Southcentral Alaska, the Kenai River and its headwater Kenai and Skilak Lakes are known not only for their salmon but also for their brilliant blue color.

Most Alaska freshwater bodies freeze over in winter but very rarely to their bottoms. Fish and mammals like beavers and muskrats continue to live under the ice. Alaska's remarkable blackfish gather together in balls and lower their metabolism to survive along lake and river bottoms during the winter, and as the only fish in Alaska capable of breathing air—which they do through holes in the ice that they help keep open with their movements—they enhance their winter survival in low-oxygen sediments. Birds like loons and grebes that nest on lakes migrate to offshore waters for the winter months.

Spring comes with what Alaskans call breakup, when river ice breaks apart and flows downstream. It's a time of great movement and change, as riverbanks are scoured and reformed. In years when river ice breakup occurs quickly, massive ice jams can block downstream flow and cause extensive flooding.

Millions of migratory birds who winter as far south as Antarctica depend on Alaska wetlands for spring and fall migrations, as well as summer nesting. Wetlands are lowlands covered with shallow water or in saturated soil. These edge communities—transitional zones between aquatic and terrestrial habitat—include salt marshes, swamps, bogs, wet meadows, grass wetlands, sedge wetlands, muskegs, and peatlands. Alaska has more area covered by wetlands than do the other forty-nine states combined.

Aside from their importance as habitat, wetlands control flooding and recharge groundwater. Bogs, which represent many thousands of years of wetland succession, help regulate global climate by storing enormous amounts of carbon in peat. Although Alaska's Arctic Slope gets very little precipitation, its wetlands are extensive because permafrost keeps water from draining into soil.

However, with climate change and permafrost melt, wetlands throughout the state are draining and drying, and shrubs are expanding across drier land. Along with the warming climate, threats to Alaska wetlands come primarily from filling and dredging for transportation corridor construction, hydrological fragmentation, and pollution from poorly managed mine waste discharge.

Encompassing spongy wetlands, quiet ponds reflecting sky, the soft chatter of a mountain stream, and the deafening roar of a river raging with springtime melt, this tapestry of water is the lifeblood of Alaska's land and sea.

Red (Sockeye) Salmon

Oncorhynchus nerka

The Red or Sockeye Salmon is one of five species of Pacific salmon native to Alaska, all of whom have remarkable life histories. They're all anadromous; they start their lives in freshwater, travel downstream to the ocean to live their adult lives, then navigate back to their natal freshwater systems to spawn. They're all considered keystone species, playing critical roles in the health and functions of entire ecosystems.

Both male and female Red Salmon die after laying and fertilizing eggs in gravel nests called redds, usually in river channels attached to lakes or along shallow lake shorelines. Eggs hatch during the winter, and the young, known as alevins or sac fry, live in the gravel off their yolk sacs until they emerge in spring as small fry, to feed on plankton and insects. The match in timing between their yolk-sac consumption and a plankton bloom is critical in their early survival.

Depending on their home stream systems, Reds spend one to four years in freshwater before swimming downriver as finger-length smolts. They then live in the ocean for another one to three years, feeding and traveling long distances around the North Pacific before migrating back home. While at sea, Reds feed primarily on zooplankton, which they filter through long, serrated gill rakers. A mature sea-run Red is sleek and shiny-metallic, weighing an average of 5 or 6 pounds.

Once Reds reenter freshwater, their bodies turn bright red and their heads green, and males develop humped backs and hooked jaws lined with sharp teeth. The Red Salmon's names describe this transformation. *Sockeye* is an Anglicized version of the word for red fish from a Coast Salish language in British Columbia, and *Oncorhynchus* comes from the Greek for "lumpy (or hooked) snout."

Salmon life is hard. Of the two thousand to five thousand eggs a female will lay, nearly all eggs, fry, smolts, and juveniles will become food for other beings. Only a handful will survive to migrate back to their rivers. Humans, orcas, brown bears, large ocean fish, seabirds—hundreds of species depend directly on salmon throughout their life cycle. And hundreds of other species depend on them indirectly. Spawned-out carcasses play an essential role in transporting nutrients to rivers, upland areas, their own progeny, and other beings including plants and trees.

Hundreds of populations of Red Salmon exist in Alaska, each fitted to its specific river and lake system. The largest populations come from the nine

river systems that empty into Bristol Bay. In recent years, returns there have been in the forty million range, most swimming into river mouths within a one-month period. In fact, half of the world's total commercial catch of Red Salmon comes from Bristol Bay alone. The key to that abundance is primarily the near-pristine upland habitat of streams, lakes, and wetlands they rely on, combined with favorable (for them) ocean conditions.

While pink salmon are Alaska's most numerous salmon, with their numbers enhanced by hatcheries, Red Salmon are a close second and of far greater economic value. As human food, Reds are known for their firm red or orange flesh, rich flavor, and high oil content. They are the backbone of Alaska's commercial salmon fisheries but are today also highly valued by subsistence, sport, and personal use fishermen. They hold critical nutritional and cultural roles for Alaska's Indigenous Peoples, who have long traditions of honoring them. Their high nutritional content also makes them essential to the wild beings who depend on them.

Other salmon populations in Alaska, especially king (chinook) and chum, have been in sharp decline, with individuals getting smaller due primarily to ocean and river warming, but Red Salmon, at least for now, may be benefiting from environmental change. Studies have shown that warmer lake and stream water has been good for plankton growth, allowing Reds to put on more weight more quickly. Reds in Bristol Bay now usually spend only a single year in freshwater before migrating to the ocean. The picture is not simple, though. Those same fish appear to remain an extra year in the ocean, as climate stressors and competition from hatchery fish have decreased available food there.

Returning from their long lives in salt water, navigating epic migrations back to where they were born, Red Salmon link land and sea. And they're a sight to behold, whether it's their countless swaying red bodies pushing against the current of a clear-running river or their leaps up waterfalls as massive brown bears try to snatch them in midflight.

ANNIE WENSTRUP

Ggugguyni Transcribes the Archive as Diviner

What treasure can I offer the auger?
Clear water in a silver bowl. Bird bones.
Feathers. A red bead. My open hand,
palms offering copper coins, tin nails, oil
paints. The Hubble Telescope's* first picture.
Yesterday's paper, horoscope circled.
Lines. An ant's path, The magician's card
deck, ends tapered. My first memory. Mist.
This is what I learned of the future:
a red salmon, her scale-blushed belly,
her silvered cheeks above her gills.
Her gills like a door in a lift-the-flap book.
Here, hook the crook of your index
finger and pull. There's the lure.

* Earthdate 12.25.2021: On TV, a NASA scientist tells the Hubble *goodnight sweetheart, I love you.* Meanwhile, I cry. I don't know if I cry for the man, or the Hubble, or for myself, for all of us far from home. Meanwhile, a Star Trek *Voyager* rerun begins.

Tundra Swan

Cygnus columbianus

With the graceful S curve of their long necks and their brilliant white bodies gliding through emerald waters, Tundra Swans are among the most elegant of Alaska's birds. They're also large: the adult Tundra Swan weighs about 15 pounds and has a 6-foot wingspan. Their plumage is a brilliant white except on their necks, where it's often stained a rusty orange from feeding half submerged, tails in the air, on aquatic plants growing in iron-rich sediments. Feet and bills are black, with a small splash of yellow marking the space between bill and eyes. That yellow spot is the best way, other than size, to tell the Tundra Swan from the trumpeter, the larger and less common of the two Alaska swans.

The male is called a cob, the female a pen, and a young one a cygnet. Where do these odd names come from? *Cob* reportedly stems from an old English word meaning "leader," and *pen* from the fact that female swan feathers were the best for making quill pens. *Swan* itself links back to a word for sound, honoring the bird's high-pitched, quavering *oo-OO-oo* calls. The Tundra Swan was formerly known as the Whistling Swan after the noise of their wingbeats.

Alaska's Tundra Swans belong to two distinct populations. The eastern population of about ten thousand, after summering on tundra lakes in Alaska's Arctic, migrate all the way to the mid-Atlantic coast to winter. The larger western population nests along Alaska's west coast and the Alaska Peninsula and migrates to wintering grounds from southern British Columbia to central California. Some Alaska Peninsula swans do not migrate at all but spend their winters in open water along the peninsula's southern coast. Migrators can take months to travel, stopping along the way to rest and feed, spending more time on the journey than in either summer or winter locations.

All species of swans are renowned for pairing for life, though if one mate is lost the other may find a new mate. Tundra Swans spend a year together before mating. They then build nest mounds from plant material next to tundra ponds or lakes, where three to five eggs are laid. Both parents share in egg incubating and guarding the cygnets until they fledge. They are extremely territorial, chasing off not just other swans but also geese and ducks and even larger animals like wolves. They return to the same nest site every year and can live for more than twenty years.

Alaska's Indigenous Peoples traditionally hunted Tundra Swans with bows and snares for food, feathers, skins, and bones; they also gathered their large eggs. Archeological evidence from Kodiak shows that Tundra Swans' long, sturdy wing bones were fashioned into awls. Swans continue to be taken for subsistence in some parts of Alaska. While market hunting, lead shot pollution, and the destruction of southern wetlands once threatened Tundra Swan numbers and brought trumpeters close to extinction, both species have adapted by shifting to winter foraging on waste products in agricultural fields; they also may benefit from longer open water and growing seasons as a result of climate change.

Tundra Swans are devoted to family: they migrate in fall with family groups, traveling in a V formation for thousands of miles from the high Arctic to the Lower 48, where they gather by the hundreds on large inland lakes and fields, providing a winter delight.

PEGGY SHUMAKER

Swans, Where We Don't Expect Them

Tundra swans twine necks
among snowflakes
vanishing into evening's

river. Past breakup,
tablecloths of rotten ice
nest along the bank.

Halfway, swan wings
open, then settle in
like second thoughts.

Maybe they flew
north over Minto,
traced halos

over brooding ponds,
saw from far up
without touching

the world is hard
and will stay hard
a while longer.

Feltleaf Willow

Salix alaxensis

Also known as Alaska Willow (along with Big Willow, River Willow, and Uqpik in the Iñupiaq language), Feltleaf Willow is said to be Alaska's most abundant tree, although most people would call these beings shrubs. The tallest may reach 30 feet but most are much shorter, especially in the northern part of the range.

Feltleaf is one of more than forty willow species found in Alaska and is common just about everywhere in the state except for the Aleutians; this being's territory extends into Canada but not the Lower 48. As the climate warms, willows are moving northward and growing taller. One recent study found that Feltleaf Willows on the North Slope are increasingly available to the moose and snowshoe hares that are moving northward with them.

True to their name, Feltleaf Willows are known for the soft, feltlike undersides of their pointed, boat-shaped leaves. Because they germinate quickly, they spread easily in disturbed areas—along rivers and streams, in areas exposed by melting glaciers, and where people have cleared for homes, roads, and parking areas. They are also fire- and flood-adapted and can resprout after being burned, submerged in water, or washed downstream. They reproduce from furry catkins (sometimes called "pussy willows") as well as by resprouting from broken stems. Such stems root so easily that a common way of stabilizing eroding stream banks is to stick cut branches into the soil.

The Feltleaf and its fellow willows have been called the little engines that make the northern forest go. They are the major source of food for Alaska's moose, who eat 30 to 40 pounds of leaves, stems, and bark each day. Snowshoe

hares also eat the leaves, bark, and small twigs. Willow ptarmigan feed on both buds and leaves.

Take a walk among willows and you'll see plenty of evidence of moose and hare browse—the snipped-off ends of the reddish twigs. A single willow plant can lose 90 percent of its growth to browse and grow back quickly, as much as 3 or 4 feet per year. Biologists say willows have been "broomed" when heavily browsed by moose because the many shoots they then send out make the plants look like a broom.

In areas without large trees, Feltleaf Willows can provide fuel for cooking and warming fires. The young, tender shoots, with bark removed, can be eaten by humans and are said to taste like cucumber. The roots have long been used by Indigenous Peoples for basketmaking; in fact, the popular whale baleen baskets sold to tourists are said to be made with a technique adapted from making willow baskets. Willow bark is high in salicin, "nature's aspirin," and chewing buds or bark is a traditional way of relieving headache or other pain.

JOHN STRALEY

Two haiku

A cold gravel road
with willow leaves glittering;
Grief comes . . . anyway.

Yellow willow leaves
Perched on the beach like canoes
with no one around.

Bluet Damselfly

Enallagma cyathigerum

The very slender Bluet, at most a couple of inches long, is common throughout Alaska, all the way up into the Arctic. The male shimmers in neon blue with black bands evenly spaced along his thorax, while the female can be either blue like him or a drabber green-brown better suited for blending into her surroundings. Bluets live in and around freshwater, where most of their life cycle is spent in the larval stage. Because they thrive in unpolluted waters, they are important indicators of aquatic health.

If you've hung around wetlands, ponds, or lakes, you have likely seen Bluets coupled in what is known as a mating wheel. The male holds the female by her neck, and the female bends her body around to join with his reproductive parts. They fly in that pairing to find a suitable aquatic plant and deposit eggs there, just below the surface.

The larvae, called nymphs, resemble the adult form, with large heads and tubular bodies. They feed on aquatic insects and go through as many as a dozen molts before maturing, eventually climbing out of the water for a final molt into adulthood. A generation can span two years if the larvae linger through winter in the larval stage. Their flying forms generally live just a month or two during Alaska's short summers. In short, the reason Blue Damselflies take to the air is to mate!

Bluets are considered intermediate predators—that is, they eat smaller insects and are eaten in turn by larger insects and birds. Powerful, agile flyers, they grab prey out of the air.

While damselflies and dragonflies belong to the same scientific order, Odonata, they differ in body shape and wing position; damselflies are thin and rest with their wings folded while the stockier dragonflies spread their wings out. Combined, they number about five thousand species worldwide. About thirty of these species live in Alaska. Alaska's state insect, the four-spotted skimmer, is a dragonfly. When in watery places, listen for the hum of Odonata and watch for the speedy brilliance of Bluets.

EMILY WALL

Damselfly Hymn

In this polished morning
a damselfly, with his tender
feet, alights on still water.

He could be singing
or asking for help,
or perhaps just yearning

for clear water, clear
air, for me to be standing
not so close.

Am I in his kitchen?
Have I slipped in to help
myself? I swallow

the vision of him, drink
a tumbler of beauty at his
small, silver sink. If this was

a church, I'd say, amen.
If this was a mosque I'd remove
my shoes. Just once, I want

to leave, not hungry. Just
once, I want to be the shape

of the song.

American Beaver

Castor canadensis

American Beavers are the largest rodents native to North America, with adults weighing on average 40 to 70 pounds. Their broad tails act as rudders for swimming and as braces when they stand to cut trees. The warning sound of a tail slapping water will be familiar to anyone who's explored around a Beaver pond.

Beavers generally live in wet forested areas, including on Kodiak Island, where they were introduced in 1925. They are famous as chisel-toothed engineers and symbols of industriousness—as in "busy as a beaver." The orange color of their teeth comes from iron in the tooth enamel, which makes their teeth stronger. Their teeth continue to grow throughout their lives.

By cutting trees and building dams and canals, Beavers change the hydrology and create habitat for fish, insects, birds, plants, and greater diversity in general. They are especially vital to salmon health. By creating ponds, Beaver

dams enhance rearing and overwintering habitat that shelters juvenile salmon from high-flow events. Beavers increase water storage and decrease water temperature in headwater streams. They are less popular with humans when they block culverts or flood properties.

Alaska's Indigenous Peoples—perhaps especially the Athabascans, with whom Beavers have lived closely—respect Beavers not only for their meat and fur but also for their powerful, sensitive spirits. In an origin story of the Haida from Southeast Alaska, Beavers are descended from Beaver-Woman, who built a dam on a stream and gave birth to the first Beavers.

These beings were once aggressively trapped for their thick furs, but with the decline in the fur market their numbers have been increasing. Their range has also been expanding as they move higher up into mountain valleys and into Arctic tundra, following the northern movement of shrubs and trees as the climate has warmed. Aerial photos from the 1950s show no Beaver ponds in the Arctic's tundra. Recent satellite images show more than eleven thousand Beaver ponds on the same lands. In 2002 there were two Beaver dams near Kotzebue; by 2020 there were ninety-eight. What had been streams have been reengineered into necklaces of ponds. Implications of the northward spread are not clear; two questions yet to be answered are whether the new ponds in the Arctic may be thawing permafrost beneath them and whether fish migrations may be impeded.

Beavers live in families of parents, yearlings, and two to four kits in a den either dug into a stream bank or in a domed lodge in a pond; either structure is covered with branches, sticks, and mud. Each den is spacious enough to serve as a sleeping and rearing area, food cache, and general year-round home for the family. Key to their survival is water deep enough to remain unfrozen throughout the year so that the Beavers can come and go through their tunnels to secure the bark, roots, leaves, and aquatic plants they store outside the den for their food. Their dams, built of logs, sticks, mud, grassy debris, and rocks, are raised to increase water depth. In very cold winters, if ponds freeze to the bottom, Beavers denning in them will die.

Young Beavers stay with their parents for two years and then leave to find mates and new homes, traveling as much as a dozen miles to find suitable habitat. You may do a double take when spotting one on a beach or in salt water on the way to a new territory.

JO GOING

Friendship

That beaver, his house
too close to the trail, too near footprints.
Each morning, there he is, felling, carrying, building . . .
I know him.
Before, he would slap the lake with that curious tail,
then dive.
Now he looks at me looking at him, and, stick in mouth, continues storing.
I am told of a tough old miner who hung a sign by the river, his lost words
now my prayer:
PLEEZ DUNT SHOOT THE BEEVERS THAY AR
MY FRENDS.

Sandhill Crane

Antigone canadensis

Who doesn't marvel at the sight of these large prehistoric-looking birds flying overhead with legs trailing behind them, and at their bugling calls? In Alaska, the arrival of Sandhill Cranes in the spring is a sure sign of a new green year.

Sandhill Cranes do, in fact, bear a more than superficial resemblance to dinosaurs. They have the longest fossil history of any now-living bird, going back at least 2.5 million years. (A 10-million-year-old crane fossil may belong to a relative or ancestral crane, not the Sandhill specifically.)

This elegant, large-bodied being stands about 3 feet tall, with long legs, a long neck, and a wingspan of 6 feet—making them among Alaska's largest birds. Those who breed in Alaska, called Lesser Sandhills, are smaller than the Greater Sandhills, who breed in more southerly locations. Plumage is gray, although some nesting birds dip their dagger-like bills into muddy ground and "paint" their feathers with iron-rich mud, presumably as camouflage. Drooping feathers over the tail form what looks like a fancy-dress bustle. The adult bird's crown—which is skin, not feathers—is a bright crimson red.

These birds get their name from the Nebraska sandhills along the Platte River, where they gather in huge flocks during their spring migration. The Yup'ik name, Qut'raaq (along the Yukon River) or Quciligaq (along the coast), comes from the bugling sound of the call. The Iñupiaq name is Tatirgaq. The *Antigone* part of the Latin name comes from the Greek myth in which Princess Antigone (daughter and half sister of Oedipus) was transformed into a crane.

Sandhill courting behavior is especially dramatic, as the birds lift their wings, pump their heads, bow to one another, leap high, and sometimes throw sticks or grass into the air in a surprisingly strange and marvelous choreography that ends with a deep bow. After mating, pairs (who stay together year after year) nest across Alaska's wet tundra, muskegs, and marshes on mounds of grasses and sedges. The females typically lay two eggs, and both adults take turns incubating them for a month. The fluffy yellow young, called colts, are well developed and active once hatched.

Sandhills are omnivores, eating seeds, bulbs, berries, insects and other invertebrates, even smaller birds and rodents. They fiercely defend their nests and young from predators like eagles and coyotes. In general, Sandhill Cranes are considered abundant, with their populations increasing about 4 percent each

year between 1966 and 2019. This increase is largely due to their adaptation to winter feeding in agricultural fields.

Two distinct groups of Lesser Sandhills live in Alaska, coming and going over different migration routes. About twenty-five thousand fly along the Pacific Flyway to and from the Central Valley of California to breed in Alaska's Cook Inlet and Bristol Bay regions. Perhaps half a million that winter in Texas and the Southwest fly over the middle of the continent to nest in western and northern Alaska as well as across the Bering Strait in Siberia.

Sandhills typically arrive in Alaska in late April and early May and leave in August or September. Family groups stay together and return together in the spring, although juveniles then separate from the parents to flock with other nonbreeders. After two years, juveniles find their own mates. Individual birds can live as long as twenty years in the wild. (One in a zoo lived sixty-one years.)

One special place to watch Sandhills is at Creamer's Field in Fairbanks, where several thousand birds stop to refuel during their migrations. The Tanana Valley Sandhill Crane Festival is held there each August, when the Sandhills stop on their way south. Their loud bugling overhead, which can be heard more than 2 miles away, announces their return.

CAROLYN KREMERS

Crane Song

O sandhill cranes
lifting from the fields

please carry
north and south

among your strong
and ancient cries

the letter
I never wrote

the words
I would have said

could I have flown
to see my father

one last time

Arctic Grayling

Thymallus arcticus

The unique beauty of Arctic Graylings derives from their large dorsal fin, resembling a sail fringed with red and dotted with spots. The purpose of these fins—taller and longer on males than females—is thought to be intimidating rivals, deterring predators, and attracting mates; in mating, the male holds this fin against the female during the release of eggs and milt. The largest Arctic Graylings can grow to a length of 20 or so inches and weigh 5 pounds. And their color doesn't stop with the dorsal fins: the sides of individual Graylings shimmer in silver, gold, or blue, changing color to match their rivers.

Graylings belong to the same family (Salmonidae) as salmon and trout and are associated with clear, cold rivers and lakes. One of six species in the northern hemisphere, the Arctic Grayling is the only one to live in Alaska, where the native range covers the whole state except for Southeast, Kodiak, and the Aleutian Islands. The entire range includes Arctic Ocean drainages in Canada and Russia. Arctic Graylings once lived as far south as Michigan and Montana but now are very uncommon outside of Alaska (except where artificially stocked) due to overfishing, habitat loss, and competition from introduced species. In Alaska, they're eagerly sought by sport fishermen as well as serving important subsistence needs.

Graylings are voracious and omnivorous eaters: they feed on aquatic and airborne insects, crustaceans, salmon eggs and smolt, sometimes even on small voles or shrews that fall into the water. During peak mosquito season, they gorge on mosquitos. They're an important link in the Arctic food chain, as they not only help keep mosquito populations in check but also serve as food for other beings like pike, sheefish, eagles, osprey, river otters, and mink. In winter Arctic Graylings feed minimally while living under ice, where they tolerate low oxygen levels that would kill other fish.

According to some sources, an individual Grayling can live as long as thirty-two years, spawning multiple times. Unlike salmon, they don't excavate nests; instead, during their vigorous courtship they tend to stir up bottom material that covers and sticks to the adhesive eggs. Within a year of hatching, the young fish migrate upstream and downstream—sometimes short distances along a stretch of river, sometimes for a hundred miles—to seek their ideal areas for feeding, spawning, and overwintering.

The *Thymallus* part of the Grayling's Latin name, given by Linnaeus back in the eighteenth century, refers to the smell of thyme, thought to emanate faintly from the flesh. Next time you meet an Arctic Grayling, give a sniff.

FRANK SOOS

The Blue Fish

In a land where every spring
Makes this river new,
Few things grow old.
The blue fish, its body, once a
Narrow gray torpedo, transformed—
Indigo, calico, the many
Ways blue is blue—changed by
Time alone, its turquoise-splashed
Dorsal tag, its pectoral fins, shot through
With rays of black and orange.

We'd like to think it got that way
Through its fishy wisdom, but, really,
It got that way just by living.

The first fish I ever caught
Here was a blue fish. I held it in my hand,
Wonder-struck, then let it go—I didn't
Know.

I come here often. I come back,
Hoping to catch another.

Braided River

With their ever-changing patterns of sinuous waters joining, separating, and rejoining, Braided Rivers are among Alaska's most iconic natural features. Unlike older mountain ranges such as the Appalachians, Alaska's fourteen major mountain ranges have so recently been glaciated and shoved skyward by volcanic and tectonic activity that their valleys are U shaped rather than V shaped. In these broad valleys lie the Braided Rivers: meandering interwoven strands of water divided by ephemeral islands called braid bars.

Braided Rivers wind throughout Alaska and vary greatly in width, length, and drainage. The McKinley River in Denali National Park, draining the north slope of the Alaska Range, stretches over a mile wide. The 290-mile-long Copper River, draining the Wrangell and eastern Chugach Mountains, is 50 miles across at the biologically rich braided delta before entering the Gulf of Alaska. And the 2,000-mile-long Yukon River, stretching from Canada to the Bering Sea, winds and braids as an important transportation corridor for people and other beings.

Braided Rivers are created by the wide valleys and abundant sediment of young, eroding mountains and by the pulsing, rapidly changing flow from glaciers, icefields, rainfall, and snowmelt. As their valleys broaden and flatten out, their waters lose velocity and their sediment drops out to form wide gravel floodplains. They constantly shift, especially during times of high water, carving new channels and braid bars. Their meandering undercuts banks to sweep trees into the river, creating logjams; these eroding banks sometimes uncover ancient fossils, even remains of oxen and woolly mammoths. On a sunny midsummer day, water levels can rise so rapidly that braid bars are covered and channels become dangerous to cross. Many channels, however, are usually fairly shallow. And even at high levels, these rivers never fill from bank to bank.

The sources and sediment loads of Braided Rivers determine the color of their waters, which can range from crystal clear to cerulean to gray-blue with silty glacial melt so dense the depth is deceiving. The 22-mile Resurrection River, pouring from the Harding Icefield's Exit and Lowell Glaciers, roils with gray glacial silt. The Yukon River's name derives from the G'wichin phrase chųų gąįį, meaning "white water river," referencing not the waves of rough water but the millions of tons of glacial sediment scoured from the world's most massive nonpolar icefields upriver.

Most braid bars are bereft of plant life, but alder, willow, and many other plants crowd the banks and sprout on larger, more stable islands. The broad, threading waterways and their clustered vegetation host a diversity of insects and are magnets for birds and mammals. Moose, bear, caribou, wolf, and others can be spotted walking Braided River beds in search of food and shelter. And salmon easily navigate the braids to their spawning grounds, as they have for millennia. With graceful beauty, Braided Rivers scroll messages of time's passage across the landscape.

KAYLENE JOHNSON-SULLIVAN

Killik River, Gates of the Arctic

The river trip began with hot sun and low water.
For two days we dragged the raft over shallows

and riffles, straining and sweating. On day three
we floated past the smell of sulfur,

cool caves melting permafrost. No one spoke
as we listened to the drip of melting ice.

Tributaries began pouring into widening
channels, the water deepened, and

our shadows sped along the river's banks.
Soon a torrent, the river jumped its path and

raced through tall willows bent low beneath the
force of water. We ducked, paddled, bounced off

vegetation, including alder which no one
could remember seeing this far north.

Branches of a submerged tree caught the bottom of
the raft. Bucking against the wild current,

we threw our weight high side, rocked
until the tangle released us.

The river spat us out
onto woven gravel bars, littered with

uprooted trees. Later at camp, someone said,
"I've never paddled through a forest before."

We might have laughed except
the whispers of these warming Arctic waters

are a song of lament. That night
we heard the howl of wolves.

Round-leaved Sundew

Drosera rotundifolia

Pressed to the ground, tucked among sphagnum, the Round-leaved Sundew is easy to miss. But a closer look reveals the rosette of paddle-shaped green leaves covered with bright red hairs, each hair tipped with a glistening droplet that looks like dew in sunshine. It's this dewdrop appearance that inspired this being's common name. And it's these little beads of sweetness that attract

insects, who become stuck and then trapped as other hairs nearby bend in, completing the capture.

Also known as Common Sundew, the Round-leaved Sundew is one of Alaska's eight species of carnivorous plants—plants that consume insects as a dietary supplement. Alaska also has English sundew, two species of butterwort, three species of bladderwort (an underwater carnivore), and one species recognized as carnivorous only in 2021—western false asphodel. Many botanists agree that more of Alaska's plants may be carnivorous.

Round-leaved Sundews are circumboreal, growing throughout the northern portion of the northern hemisphere. They're one of the most widely distributed of the two hundred species of sundews worldwide and grow as far south as Hawaii. In Alaska, Round-leaved Sundews grow only a couple of inches across and flat against the ground, but when they bloom in summer, small white-pink flowers rise on stalks up to 9 inches tall. They self-pollinate, so that alluring nectar is only for carnivory. A tenacious plant, the Sundew survives Alaska's cold winters by forming a resting bud of tightly curled leaves at ground level; this is called a hibernaculum.

Round-leaved Sundews depend on open spaces with plenty of light and thrive in nutrient-poor, acidic soils. They're found in muskegs, fens, bogs, and wet stands of black spruce throughout Alaska—basically anywhere that's wet and open. They're particularly fond of growing in association with sphagnum moss, perhaps because sphagnum can pull water up from deep below the surface and can hold forty times their weight in water, providing moisture even in the driest of times.

If you walk on damp ground, take care and watch for the many small, unassuming, yet fragile plants tucked away, including Round-leaved Sundew, with tentacles radiating from the center like a red-green star, beckoning.

MARYBETH HOLLEMAN

Sundew: from that which appears inconsequential

here you are:

tiny rosette of red

hugging this wet mound

of ground. here: where

I have learned

to find you: down

low where the winged

might land, wanting

your sweet nec-

tar to sip but stick

they do, stuck,

and your ten-

tacles hold tight,

so small and dewy

like morning's

first soft and

harmless light.

King (Chinook) Salmon

Oncorhynchus tshawytscha

King Salmon are widely regarded as among the most magnificent fishes in Alaska, not only for their large size but also for their extensive migrations, astonishing life cycles, and value to other beings in their habitat. *Oncorhynchus* is a genus grouping all Pacific salmon and translates from the Greek as "lumpy (or hooked) snout." *Tshawytscha* comes from a Native Siberian word for this large fish. Alaskans usually call them Kings, although elsewhere in the country they're more commonly referred to as Chinooks. Other names include Blackmouth (for their black gums), Spring Salmon, and Tyee Salmon.

King Salmon are the least abundant and the largest of the five Pacific salmon found in Alaska, with adults often exceeding 30 pounds. (The other four are the red or sockeye salmon, the coho or silver salmon, the chum or

dog salmon, and the pink or humpy salmon.) The largest King on record, caught in 1949 in a commercial fish trap in Southeast Alaska, weighed 126 pounds; the largest caught by rod and reel, from the Kenai River in 1985, weighed more than 97 pounds.

Like all Pacific salmon, Kings are anadromous; they hatch from eggs in freshwater, spend a year in a river channel, migrate to the ocean to feed and grow for one to five years, then return to their same river to spawn and die. As large, strong fish with impressive fat reserves, Kings are built for long river migrations, bringing them far into and through Alaska's interior. On their journeys up the Yukon River, they may weigh as much as 90 pounds and travel up to 2,000 miles to reach spawning grounds in Canada.

How Kings and other salmon find their way back home is still something of a mystery. Scientists suggest salmon navigate in the open ocean by using an inherited map of Earth's magnetic field like a compass, enabling them to locate rich offshore feeding grounds. This "compass" helps them return to the near-shore region of their birth, and then they use an incredibly fine-tuned sense of smell to find their natal home stream. They build this smell memory bank as juveniles, before they start their migration to the ocean.

King Salmon are a vital part of their habitats, a keystone species. Because of their early-season arrival, energy-packed fat, and large size, they're necessary food sources for many species, including orcas, seals, bears, and large birds of prey. Resident orcas are especially dependent on King Salmon. After death, Kings provide key nutrients for streams, rivers, and forests.

King Salmon have always been the most highly valued of Alaska's salmon for human sport, commercial, and subsistence fishing. Their value nutritionally and culturally to Alaskans is of singular importance, and they're deeply entwined in the lives of Alaska's Indigenous Peoples as an essential food source. These peoples (and others) have used both nets and fish wheels (powered by river currents to scoop fish into boxes) to capture Kings from the rivers. For preservation, the fish are typically cut into strips and dried or smoked, with techniques passed down through communities and families. Alaskans typically celebrate the King's yearly return with gatherings that acknowledge this iconic being's role in renewal, abundance, and dependability.

In recent years the numbers of King Salmon throughout the state have been in sharp decline, probably in response to environmental changes including ocean conditions. Like all of Alaska's salmon species, Kings are getting smaller. Their long migrations also make them vulnerable to many interceptions along the way and make their management for sustainability difficult. Fishing closures, even for those humans with deep generational use patterns and dependence, have become increasingly necessary for conservation.

CARRIE AYAGADUK OJANEN

the soul is a

salmon

is the sun-
series on the sea
a thousand
suns

each upwelling is the soul

again
again

look
how the sea is today—so gentle

she sucks the pebbles

softly

lips round beads loosely
the sun shines the smooth rocks

wetted in her mouth

I will lift each salmon

I will lift them up
I will heft each soul
I will hold it up
chest high

the salmon's scales glint on my fingers each scale a sun

how beautiful is this salmon, this salmon I hold in my hands
see its body, net-wrenched as it is
perfect

*****

this is how to unfold flesh—

we eat together, on the floor
the red tablecloth beneath us.
the enamel pans filled full—dense
brown meat, pink flesh,
salmon egg sacks, coffee cans
shine with seal oil, gleam
green with *syrah*, our hands
reach down—the flesh flaking
in our fingers.
she is smiling
my Aaka, laughing,
her dark eyes gleaming.

Aġuvitisi.
Sit down with us.

Syramic. The seal oil is passed, poured out,

her fingers glisten
how brown her fingers
how lovely the deep blue veins

Arctic Tern

Sterna paradisaea

Nothing heralds the return of summer's midnight sun like the graceful, buoyant flight of Arctic Terns. Arriving on their Alaskan summer grounds in early spring, they're so small and slender it's hard to imagine how they manage one of the great wonders of the natural world, a migration longer than any other being on Earth.

These lissome water birds migrate more than 50,000 miles a year, making a twice-yearly journey from wintering grounds on the edge of pack ice in Antarctica to nesting grounds in the Far North and Arctic. This epic migration happens almost entirely out of sight of land. And given their long lives (some live more than thirty years), they may travel 1.5 million miles in a lifetime, the equivalent of three round-trips between Earth and the Moon. With a continuous worldwide circumpolar breeding distribution, they also have the largest breeding range of any Alaskan water bird, nesting from Point Barrow through the Southeast Panhandle and all watery points in between.

With their gray and white colorations and black-capped heads, they look like slimmer, more nimble versions of gulls. But their forked tails, long angular wings, and red bills and legs make them easily distinguishable. Arctic Terns are among the three tern species who summer in Alaska: Aleutian terns breed only in Alaska and Siberia and are the same size but have different markings, while Caspian terns breed in only a few places in Alaska, notably Yakutat and the Copper River Delta, and are much larger and more robust.

It's the thin, streamlined shape of Arctic Terns that helps them fly so far. They're made for migration, preferring to glide in the air for most of the year. They're so lightweight—just 3 to 4 ounces—that ocean breezes can carry them great distances without requiring them to expend much energy flapping their wings. They can even sleep while gliding. What's even more impressive is that they don't make a beeline for home but instead fly thousands of miles out of their way seeking favorable weather and food.

At their nesting grounds, courtship displays begin with males performing a ceremonial fish flight: a male passes low over a female with a small fish in his mouth, then lands and offers it to her. If she accepts, she'll join him in a high-climbing fluttering flight. Gregarious, these birds breed in large, raucous colonies near water, on the rocky or sandy ground of beaches, small islands, estuaries, lakes, and tundra wetlands. They're talkative birds, with high-pitched, shrill calls day and night. Even their chicks make peeping sounds while still in the egg.

The small, delicate stature of Arctic Terns belies their ability to defend their nestlings from predators. The most aggressive of the more than forty species of terns, they can successfully fend off any threat, whether from raptors, polar bears, or their primary natural predator, the arctic fox. Because their nests are almost impossible to spot unless they contain eggs (lightly speckled and themselves camouflaged), many an unsuspecting human walking a shoreline has suddenly been dive-bombed, loudly and vigorously, by Arctic Terns, whose beaks are sharp enough to draw blood. They also defecate on intruders, which is how they earned the Yup'ik name Teqirayuli: "the dear little bird that is good at using its bottom to disadvantage others." Birds like king eiders, red-necked phalaropes, ruddy turnstones, and the much-less-common Aleutian terns often choose to nest nearby to benefit from their fierce defenses.

Arctic Terns are also expert fishers and primarily eat immature herring, cod, sand lance, and capelin. Foraging over streams, lakes, estuaries, and the ocean, Arctic Terns have a rare ability among birds: they can, like hummingbirds, hover. They dance their quick, upbeat flight over water, then stop and hover, drop into a shallow plunge-dive, rise up and into the air with a shake, and quickly swallow the fish headfirst. Quickness is imperative, as food pirates like jaegers and gulls often chase them to steal their food. Depending on what's available near nesting colonies, Arctic Terns also eat crustaceans, marine worms, berries, and insects. In winter, they rest and feed along the Antarctic coasts.

In the late nineteenth century, Arctic Terns were among bird species decimated by the feather trade for women's hats. But with the Migratory Bird Treaty Act of 1918, most of their populations recovered. Being cosmopolitan travelers dependent on land and sea, they still contend with a host of threats: human disturbance at colonies, degradation of barrier beach and island nesting habitat, pesticides, predation by cats, overfishing of their prey species, marine debris, and habitat loss from climate change. In this century, Arctic Terns are projected to lose 20 to 50 percent of their habitat due to the warming climate.

Living in fragile balance with their watery world, Arctic Terns offer a striking reminder of how everything is connected to everything else. And since they summer in the north and winter in the south, they live in an endless summer. No other animal on Earth sees more daylight than these slender, graceful sky dancers.

ANNE CORAY

Sterna paradisaea

I love the tern, when it sweeps, china-white
against the ridge of variegated green,
its way of flashing through
once in a random evening,
glide more elegant than a gull's,
tail an open shears
arrested just before the act of cutting.

And when the bird is gone,
we cannot stay forever watching
the rose-gray light and the lake, whose color
—mountain-blue in early summer—
is turquoise-emerald by July, as silt
moves down from the upper glaciers.

We're coming close when a thing that's passed
becomes a sheen for which we cease to yearn,
and we're ready to welcome night's granite cast.

Alaska Moose

Alces alces gigas

Standing taller than a human with racks that can span 6 feet, the Alaska Moose is a majestic icon of Alaska's northern forests and their waterways. The largest member of the deer family, the Moose is Alaska's official state land mammal. (The bowhead whale is the state marine mammal.) Although Moose are found throughout the northern forests of North America, Russia, and Europe (where they're called elk), the largest thrive in Alaska, where the males can weigh up to 1,600 pounds.

Only male Moose grow antlers, which they use to joust with one another during the rut and shed later in the fall. The antlers can grow an inch a day, and the largest racks grow quite heavy, up to 75 pounds. Mating takes place in the fall, when these usually solitary animals gather in groups. Females give birth in late spring to one or two calves the color of shiny new pennies. One interesting feature in both sexes is the bell or dewlap, a flap of skin that hangs under the chin. Its purpose is unknown, but biologists speculate it may play a role in sexual selection as a visual or olfactory signal. Dewlaps in other species have been linked to the regulation of body heat.

The word *moose* comes from the Algonquian language of eastern Canada and means "twig eater." Tuntuvak, meaning "big caribou," is the Yup'ik name. Dineega is the general name for Moose in the Athabascan language, although Athabascans have an extensive vocabulary to identify the different ages, genders, sizes, and circumstances of Moose as well as every part of a Moose's body.

The Alaska Moose has always been nutritionally, culturally, and spiritually valuable to Alaska's Indigenous Peoples, in many parts of the state rivaling only salmon as the most essential subsistence food. These beings have also served as a source of hides, bones, and antlers for clothing, shelter, tools, and weapons. Human hunters take several thousand Alaska Moose each year, and they are also important prey species for bears and wolves. Moose remains are vital food for ravens, small mammals, and other birds of prey.

Moose eat enormous amounts of vegetation, around 73 pounds of food a day in summer. In spring and summer they can be seen in urban areas feeding in gardens or on their knees grazing on lawns. Their summer food is typically flowering plants, pond vegetation, and the leaves of birches, willows, and aspens. In winter they rely on the twigs of those same plants. Winter survival can be difficult, particularly with deep snow that requires more energy to move through and covers much of their food sources.

Moose most often live along river systems; they also do well just above tree line and in burned-over areas where new willow growth provides a bounty of food. As the climate warms, Alaska Moose have expanded their territory westward and northward across Alaska, following the range expansion of their woody shrub browse, and they can now be found almost everywhere in the state. The Togiak National Wildlife Refuge in western Alaska has seen a four-hundred-fold increase in Moose since the early 1990s, and in recent years Moose have traveled along Arctic rivers as far as Utqiagvik on the North Slope. Current estimates are that there are about two hundred thousand Moose in Alaska.

Moose tend to live well alongside humans, and residents and visitors alike enjoy watching these massive long-legged animals, especially in spring when they're accompanied by calves. About fifteen hundred live in and around Anchorage and will frequent yards and downtown streets. They should not be

approached; their antlers and hooves can inflict serious injuries. However, their greatest danger to, and from, humans is when they're hit by motor vehicles. Moose will often avoid deep snow by taking to packed trails, roads, and railroad tracks.

Watching such a large mammal standing in a still pond, contently eating the smallest of greens, then lying down on the bank for hours at a time, engenders a sense of wonder at their gentle, quiet existence in the expansiveness around them.

WENDY ERD

A Form of Prayer

Under a shimmer of birch
a pattern of moose
bones on a low hill
signal winter claimed all
but the signature of her passing.

She holds her unborn still
in the white ark of her pelvis,
curl of spring moss on tiny scapulae,
nagoonberries embrace her twins
in a green upwelling, a hymn.

Let me surrender
with this same grace
to a process as inevitable
as snow covering starvation
or berries ripening
against bone.

Western Skunk Cabbage

Lysichiton americanus

This bright yellow being pushes through wet ground in early spring from Cook Inlet and Kodiak south, with the flowering part ahead of the leaves. Some Alaskans prefer the name Swamp Cabbage or Swamp Lantern because the smell is not disagreeably skunky but pleasantly associated with the revitalization of spring. The scent attracts pollinators like flies and beetles. Bears and deer welcome

Skunk Cabbage as an early spring food. (Bears reportedly eat it as a laxative when they first leave their dens.)

Skunk Cabbage has a remarkable ability to generate heat by breaking down starch stored in the roots. This not only allows the plant to emerge from frozen ground but also protects the flower bud—which can warm up to 70 degrees F and melt the snow around it—from spring frosts and sends out the flower scent, attracting early pollinators. The sheath, or spathe, wrapping around the central flower spike (inspiration for the Latin name *Lysichiton*, "loose tunic") acts as an insulator. As the air temperature rises, thick leaves that can be up to 4 feet long unfurl from the plant's base.

The *Tlingit Dictionary* includes not just the word for Skunk Cabbage (X'áal') but also other words specific to its use, including áat'l for "pit lined with skunk cabbage used for food storage." According to the Native American Ethnobotany Database, Skunk Cabbage has 111 different uses. The large leaves are used like waxed paper to line berry baskets or steaming pits. The roots and rhizomes (underground stems that produce both roots and shoots) are used in teas and tinctures to treat ailments including coughs, skin sores, headaches, and infections. Most sources discourage food use because of toxins in the leaves, which can seriously burn and inflame human tissues.

Aside from Skunk Cabbage's practical uses, most Alaskans are simply delighted to welcome this bright, cheerful being—a tall, sturdy lantern set in snow—as evidence that spring is on its way.

NORA MARKS KEIXWNÉI DAUENHAUER

Spring

Skunk cabbage
punching through the ground,
punching to spring.

Arctic Loon

Gavia arctica

Who doesn't listen for and react to the distinctive wailing calls of loons? On the lakes where they nest, their summertime voices are welcome reminders that we share wild places and need clean air, clean water, and healthy natural surroundings.

Alaska is home to all five of the world's loon species. The Arctic Loon (sometimes called the Black-throated Loon) is a northern Eurasian bird that breeds mostly in Siberia, though a small number breed in northwest Alaska. The other four are the Pacific loon, the common loon, the yellow-billed loon, and the red-throated loon. The Arctic and Pacific loons look most alike (and were considered the same species, together called Arctic Loon, until 1998). Male and female loons have the same markings, and in winter all loons fade to duller brown, with white undersides. All winter in marine waters. In all seasons, they feed largely on small fish but also include aquatic insects, crustaceans, and mollusks in their diets.

All loons have heavy bodies and sit low in the water. They spend most of their lives on water, as the position of their legs far back on their bodies makes them nearly unable to walk on land. They're great divers—diving up to 250 feet deep and staying underwater for more than a minute—but need to "run" along open water before they're able to launch into the air. Depending on their size and wind conditions, loons need anywhere from 30 feet to a quarter of a mile for takeoff and landing. They land on water chest first, sliding to a stop.

Loons nest on piled vegetation at the edges of lakes and ponds. The larger the loon, the more open water they need, so larger loons like the common and yellow-billed must nest on larger lakes. Pairs mate for life and return to the same areas each year. Both parents incubate the eggs, of which there are usually two, and care for the young. Very young chicks will ride on their parents' backs, tucked between the wings.

The biggest threat to loons is egg predation by foxes, gulls, and jaegers. When they have human neighbors, they are threatened by shoreline developments and disturbances, including jet ski and boat wakes that flood nests and kill eggs. Climate change, which dries and drains lakes, is an increasing threat.

Loons of all kinds hold significant meaning in Indigenous belief systems, including those of Alaska. Yup'ik scholar Angayuqaq Oscar Kawagley wrote of the loon as a spiritual being and model of behavior, with love for life, environment, and creator. In "The Cry of the Loon," Kawagley wrote, "The loon's cry is remembering a place that was harmonious, full of beauty and diversity that Nature so loves."

QAĠĠUN CHELSEY ZIBELL

Malġi Asks Forgiveness

In a time that no one can remember,
Loon and Raven resided together on a lake,
their feathers glistening white in the summer sun.

Remember dear, the time you painted me? On a day so still and quiet.
Wouldn't we be lovely,
I thought,

wouldn't they all want to look at us?

So beautifully, you drew your lines and dashes
on my plumage.
Such care you gave.
From neck, to wings, to back. So even and perfect.

Remember dear, the time I painted you? The perfection I wanted to return.
Oh, the symmetry I wanted to draw
on you.

Forgive me, I could not. Every hopeful seam,
every line emerging
from the charcoal in my hand disappointed me.

The visions in my isuma* were failed by my hesitations.

No, it was the wind,
or a splashing fish
that left every stroke shaken. I needed to start over.
Over.
Over.

Where have you gone? Dear, you did not need to heave the ashes at me.

With a mote of a chance,

I would have fixed you.

From that time on,
Tulugaq has been all black,
While Malġi is adorned with intricate designs

and ash on the back of his head.

*Isuma: Northern Iñupiaq for "mind, thought"

Scoter

Melanitta species

Alaska is home to fourteen species of sea ducks, diving ducks that breed mostly in the north and west of the state and migrate to inshore marine wintering grounds. Three of these sea duck species are Scoters: Surf Scoter (*Melanitta perspicillata*), White-winged Scoter (*Melanitta deglandi*), and Black (or Common) Scoter (*Melanitta nigra*). Alaska supports 100 percent of the US breeding populations of all three species of Scoter, as well as all three eider species and long-tailed ducks. Threats include climate change, contaminants, and disruption in breeding areas.

Each of the three Scoters breeds on the ground in different areas near lakes or streams, in nests lined with grasses and down. In winter they form large, sometimes mixed flocks along the coasts, often in long lines, almost in single file. Other overwintering sea ducks like harlequin ducks and red-breasted mergansers may gather nearby.

From a distance, the three Scoter species are hard to distinguish and might all be known simply as black ducks. In fact, the genus name *Melanitta* is from the

Greek for "black duck." The male Black Scoter is, per the name, entirely black except for a bright yellow knob at the base of the bill. The male Surf Scoter is mostly black, with white on the back of the neck and forehead and a large orange bill. The male White-winged Scoter has a white patch on the wing and a white comma around the eye. The females of all species are brown.

The Black Scoter, the least studied of all sea ducks, is the most vocal of the Scoters. Even in winter, the males whistle long, plaintive croons. Because of their small wing size, they require a water runway to take off, and their wings make a slapping sound against the surface as they gain speed. When in flight, their wings make a whistling sound.

On their breeding grounds, Scoters feed mainly on aquatic insects, together with pondweed. In winter, they surf the waves and dive for food, mostly mussels, clams, and other invertebrates. These birds can swallow several dozen whole mussels in one feeding period; one Surf Scoter was found to have eleven hundred small mussels stored in her gizzard! In spring, they gobble up herring eggs.

Watching a flock of feeding Scoters can be mesmerizing, because these sea ducks dive in a synchronized sequence. Within seconds of each other, dozens or even hundreds of ducks will suddenly plunge underwater. When flocks are very large, one edge of the flock dives first, and then rows disappear like falling dominoes. Twenty or thirty seconds later, in the same quick, seconds-long succession, they pop back up to the surface, rafts of black birds bobbing together in the water.

X̲'UNEI LANCE TWITCHELL

T'ooch' Gáaxw (Black Duck)

Uncle likes to take a ride and watch the black ducks,
they roll in the waves in little packs and his mind
calculates their location, speaks to him about weather
and the location of salmon.

These fishermen and their habits,
he smiles at the partnership and longs for a boat ride.

We drive down the road, which in Lingít is a river,
and he tells me about his latest ideas that his sharp mind
has turned over and over during his days:

forgiveness,

I am always thankful when someone gives me
an opportunity to practice, he tells me,
as the wind howls I watch these seabirds, these ducks,
and I wonder who else sees them the way he does.

The day stretches out before us,
and the old fisherman smiles out at the sea,
their partnership extending into stormy relaxation.

Tundra

Tundra is a treeless terrestrial biome common to Arctic and subarctic regions. In Alaska, there are three distinct but related tundra habitats, which cover about a third of the land. Alpine tundra occurs in mountainous areas above tree line, including Southcentral Alaska's coastal mountains. Maritime tundra is found along the coast of southwestern Alaska and across the 1,000-mile-long Aleutian Islands, where the Bering Sea's cold waters produce a cool, damp, and windy climate. And finally, Alaska's most expansive tundra habitat is the Arctic tundra, found generally north of the Arctic Circle, where soil is underlaid with continuous permafrost.

The word *tundra* comes from a Russian word that translates as "treeless plain" and was adopted from an Indigenous Sami language. Tundra vegetation is dominated by mosses, lichens, herbs, and dwarf shrubs, with most biomass concentrated within root systems.

Tundra's treeless state is the result of factors that vary among the three types of tundra. These include short growing seasons that don't allow enough time for plants to produce wood, strong winds that regularly damage plant tissues, permafrost or rock that prevents roots from reaching deeply enough to provide stability, and cold or thin soils that slow decomposition and nutrient cycling. Whereas Arctic tundra is underlaid by thick layers of peat and permafrost, maritime tundra is underlaid by volcanic soil, and alpine tundra appears with very thin, nutrient-limited soil underlaid by rock. All three tundra habitats are generally subject to extreme winter conditions, including intense cold and strong winds that blow away insulating snow cover.

Alaska's Arctic tundra gets very little precipitation as either rain or snow and is sometimes called a cold desert. Annual precipitation (measured as rain) on Alaska's North Slope is only about 4 inches, less than that of the Mojave Desert. Despite the lack of precipitation, Arctic tundra is a mosaic of wet and dry areas, with numerous wetlands and ponds; this is because permafrost, which can extend over 2,000 feet deep, acts as an impermeable barrier, trapping water close to the surface. Arctic tundra is dominated by tussocks (lumps formed by sedges and forbs) and patterned ground (polygons with a raised center surrounded by troughs, caused by freeze-thaw ice dynamics).

The upper, "active" layer of Arctic tundra that thaws in summer is dominated by nitrogen- and phosphorus-rich peat up to 18 feet deep. This soil stratum contains a tremendous amount of carbon, both at its surface and locked in permafrost. Arctic tundra, therefore, is considered an important carbon sink and has been estimated to hold 1.4 trillion tons of carbon. However, as permafrost thaws more deeply over longer and warmer growing seasons and as surface layers dry out and burn, tundra becomes a carbon source, particularly of methane, increasing atmospheric gases and thus worsening climate change.

Sometimes described as "barren," especially as it appears in winter, tundra is far from that. These habitats are home to Alaska's hardiest beings, including more than seventeen hundred plant species such as the rare Alaskan glacier buttercup (or glacier crowfoot), huge breeding populations of birds and insects, and iconic northern ungulates like caribou, musk oxen, and mountain goats. The short summer months feature an explosion of plants and insects that supports an intricate, productive ecosystem. The tundra in summer literally hums with bees, mosquitoes, and birdsong.

Fragile and slow growing, Alaska's tundra is particularly susceptible to damaging human activities. Alpine tundra is easily destroyed by recreational activities, and Arctic tundra is particularly threatened by industrial activities, including oil and gas development, road building, and mining. All three types of tundra are at risk from ecosystem-altering changes as a result of climate warming.

Encompassing high valleys that spread below icefields in craggy coastal mountains, the windswept Aleutians, and vast stretches of Arctic tussocks, Alaska's tundra ecosystems are rich with unique and irreplaceable communities of beings who have found ways to thrive in harsh conditions.

Caribou

Rangifer tarandus

In herds that flow like a single organism through high mountain passes into wide, treeless valleys laced with streams, Alaska's Caribou navigate one of the

longest terrestrial migrations and present one of the greatest spectacles of wildlife congregation on Earth. Their scientific name, which means "wandering deer of the North," is apt: Arctic herds, gathering in the tens of thousands, travel up to 2,000 miles a year and will even cross miles of sea ice, moving like vast rivers between winter and summer grounds. On the other hand, some, like the Chisana herd, don't migrate at all.

The Caribou is a relative of moose, elk, and deer. These animals move to seek food, and in summer's insect season to find refuge on higher ground where breezes keep mosquitoes and warble flies at bay. Even newborn calves quickly adapt to the wandering life: at two days old, calves can travel more than 10 miles a day. The sizes of herds vary greatly, from the several hundred thousand of the Western Arctic herd—until recently the largest Caribou herd in the world—to the four hundred Caribou of the Kenai herd.

Alaska's 750,000 Caribou occupy thirty-two distinct ranges, encompassing the sweeping Arctic tundra and the Brooks Range, the brushy Kenai River flats, the Denali foothills, and the broad Alaska Peninsula. These ranges are defined by traditional calving grounds they use every spring. The Porcupine Caribou herd, for example, calves in a narrow band of tundra along the coastal plain of the Arctic National Wildlife Refuge, where they find just the right conditions, including an insect-dispersing breeze and an abundance of sedges, flowering plants, mushrooms, and willows to eat.

A circumpolar species, Caribou are the same as those beings called reindeer in other parts of the world, but in Alaska, *caribou* refers to wild herds, while *reindeer* refers to domestic herds (initially imported from Siberia and Norway in the late nineteenth and early twentieth centuries). Weighing between 175 and 400 pounds, they're the only deer species with both males and females growing antlers. And these antlers can be heavy: while cows maintain smaller antlers, bulls, who spar with theirs during the fall rut, must carry antlers weighing as much as 30 pounds.

Equally as astonishing as their long-distance migrations are their unique adaptations to surviving Arctic winters. Their nostril bones form a spiral, increasing the surface area of their nasal passages to warm and humidify air before it reaches their lungs—an ingenious heat exchange that minimizes energy and moisture loss on cold days. Caribou also have a dual-layer coat: a soft wool underlayer is covered by outer guard hairs that have a unique air-trapping honeycomb structure. This makes their coat durable, lightweight, and warm.

And then there are their hooves, which are much larger and more flexible than any other in the deer family. In winter, Caribou hoof pads grow hair for extra warmth; in summer, their pads get spongier for tundra traction. Their flexible hooves splay to support them like snowshoes on snow and soft tundra, and provide a suction-cup-like grip on ice. Over time, the hoof edges grow sharp for even better footing on ice. The hooves' concavity makes them like shovels for digging through snow for food, and like paddles for swimming—helping Caribou swim up to 6 miles an hour, faster than any human.

Caribou have long been integral to the lives of Alaska's northern Indigenous Peoples, so much so that Native villages were traditionally built along Caribou migratory routes. The creation story of the Gwich'in tells that Gwich'in and Caribou, who they call Vadzaih, were originally one; when they separated, they remained relatives who vowed to provide for and protect each other. When the Nunamiut Iñupiat of the Brooks Range, who had moved around seasonally with Caribou migrations, established a permanent community in the 1940s, they chose a pass through which the Western Arctic and Teshekpuk herds regularly migrated: a place they call Anaktuvuk, which means "the place of Caribou droppings."

While the size of herds can swing widely over time, current downward trends in the populations of all but a few of Alaska's Caribou herds are raising deep concerns. Industrial development, including oil drilling, pipelines, mining, and road building, fragment habitat and impede Caribou movement. Climate change and increased extreme weather events, such as late snowmelts and rain-on-snow events in winter, make it difficult for Caribou to reach their primary winter food source, reindeer lichen. Increased wildfires and changes in plant growth and insect hatch are also shifting the ground beneath them.

For nearly 1.5 million years, Caribou have been exquisitely attuned to life in the Far North. The most northern herds even change the color of their eyes with the seasons: in summer, they're gold, reflecting more light out of the eyes; in winter, they're blue, scattering more light inside the eyes to give them better vision in the dark. From their amazing adaptations to their epic migrations, which create a lacework of trails across northern mountains, Caribou truly are wild symbols of the north.

JOHN MORGAN

Counting Caribou—Prudhoe Bay, Alaska

after R. Glendon Brunk

Tundra to the horizon peppered with lakes,
aswim with pintails and loons. A fox lopes by.
The sky's aslant with jaegers, rough-legged hawks.
I'm paid to tally caribou, a science guy.

Behind me a maze of pipes, pumps, drill-
pads, gravel pits, and sludge-smudged drums,
and everywhere the reek of corporate oil.
But hey, I'm here to count the herd, which comes

(if they come at all) too fast to count, then mill
when they hit the pipe. They mass and twitch, until
one coast-bound leader steps up, lifts her muzzle,
sniffs, while baffled calves and mothers nuzzle,
just as a truck roars by, blaring its god-
damn horn and they stampede away. I hate this job!

Reindeer Lichen

Cladonia rangiferina

Reindeer Lichen—also known as Reindeer Moss, Caribou Lichen, or Caribou Moss—is not a moss but a true lichen. Alaska is home to thousands of the at least eighteen thousand lichen species worldwide, with new ones being discovered regularly. Lichens are everywhere in Alaska: crustose (flat) lichens decorate rock with eccentric shapes of pale green, russet, and gold; lungwort and other foliose (leafy) lichens adorn tree trunks; and fruiticose (branched) lichens like foam and Reindeer Lichen sprout from soil and rocks in open tundra and alpine.

Lichen is not a single organism but a marriage between a fungus and one or more algae or cyanobacteria. (Understanding the nature of lichens birthed the biological term *symbiosis* in 1876 to describe the partnership of dissimilar organisms.) The white or gray-green color of Reindeer Lichen results from the algae, who contain chlorophyll and produce food, being overlaid by white fungal threads who provide moisture and keep the algae from drying out.

Low to the ground and cold hardy, Reindeer Lichen grows extensively over vast areas of Alaska, especially the treeless Arctic tundra. Each of its many branches divides into smaller branches, resembling the antlers of caribou (and their domesticated counterpart, reindeer). The common name also points to the fact that Reindeer Lichen is an exceedingly important source of food for caribou and reindeer. In winter, a caribou's diet can be 90 percent lichen. Both caribou and reindeer, as well as musk oxen, paw through thin, dry Arctic snow to reach this lichen.

The mats formed by Reindeer Lichen and neighboring plants prevent erosion, keep soils cool in summer, help regulate water tables, and, importantly for the climate crisis, store carbon. Because they're so sensitive to changes in the environment, lichens in general can be used to detect trends in their ecological communities, including the effects of changing air quality. They rely on the atmosphere instead of root systems for water and nutrients and absorb whatever is found in the air, including pollutants.

The sensitivity of lichens to radioactivity has long been a concern for the lichen-caribou-human food chain. Soviet and American atmospheric nuclear testing in the 1950s and '60s resulted in radiation being absorbed by Reindeer Lichen and making its way into humans who consumed caribou meat. The Chernobyl nuclear disaster of 1986 caused so much radioactive contamination in Russia and Northern Europe that reindeer herding and marketing were

curtailed. A 1998 study found populations of Indigenous Peoples in Canada and Alaska with radiation exposure levels much higher than those of urban Americans but concluded that the health risks of consuming caribou were low compared to the physical, social, and cultural benefits.

Reindeer Lichen, like many others, contains an acid that can inhibit other plants from crowding them out. This acidity and the plant's texture make this lichen unappealing to humans except as a survival food, but some eat the lichen-filled stomach contents (called it'irk by the Gwich'in Athabascans) of caribou they kill. A lichen tea is sometimes used to treat various ailments, a practice confirmed by modern science's discovery of lichen's antibiotic qualities.

Like many lichens, Reindeer Lichen grows extremely slowly (just ⅛ of an inch a year), so it doesn't easily recover from being trampled, overgrazed, or otherwise damaged. Industrial development and the northward movement of shrubs and forests from climate-change-induced warming are decreasing the abundance of Reindeer Lichen.

When dry, this lichen is brittle and breaks easily; when wet, the plant becomes spongy soft and supple. On summer's tundra, clusters of Reindeer Lichen interspersed with the green of all other plants can look for all the world like summer snow.

TOYA BROWN

What Future Eyes May Not See

Curls like little girls' afro ponytails
Brittle coral of the land
Grey green or cauliflower gone pale
Soft crunch cries as caribou stand
Once prevalent now frail
Deathly grip of your atmosphere's hand
Once flowing lace like a bride's new veil
Soil of your home reduced like falling sand
A beautiful carpet which human feet failed

How can we make pollution unhand
The dying continuous tale
Of our impact on your life span?

Snowy Owl

Bubo scandiacus (formerly *Nyctea scandiaca*)

Most owls hunt under cover of night, but not the Snowy Owl, who breeds in the high Arctic where the summer sun doesn't set. Instead, this shockingly white owl with piercing yellow eyes hunts anytime, especially preferring the crepuscular light of the midnight sun. With the densest and longest feathers of any owl, including feathers on their legs and feet, Snowy Owls are among Alaska's best camouflaged and most charismatic birds.

Supremely adapted to the cold, windy life above tree line, Snowy Owls perch stoically on hummocks or rocks, sitting still for hours while scanning the tundra for their favored food, lemmings. A single Snowy Owl can consume as many as sixteen hundred lemmings in one year. The lemming population's boom-and-bust cycles dictate the owl's breeding: in boom years, Snowy Owls may raise two to three times as many young; in bust years, they may forgo breeding entirely. They're highly bonded with these small 1-ounce rodents, but they'll also eat waterfowl, ptarmigan, hares, and weasels, and can even catch small birds on the fly. John James Audubon recorded a Snowy Owl crouched at the edge of an ice hole, waiting and then catching fish with those feathered feet.

Living as they do where snow often covers the ground, these beings not only have excellent eyesight but also use their incredible hearing to detect prey under snow. Like all owls, they have ears that are asymmetrically positioned to pinpoint sound sources. Then, they fly low or just run on the ground to pounce on prey. At times, they're the only conspicuous beings on the windswept Arctic tundra.

Considered among the most distinctive birds in the world for their plumage, males turn all white at maturity, while females retain some black-brown stippling. Their chicks, born in a dry depression on the tundra, transform to downy white fluff balls hours after birth. In winter, these owls migrate south as far as the northern contiguous United States, seeking wide-open, rolling, treeless terrain. Fiercely protective of both wintering and nesting grounds, they will use their 5-foot wingspan to take on animals much larger than themselves, dive-bombing humans and even attacking gray wolves.

The Yup'ik respect this owl, known as Anipak, for having strength and stealth; this is reflected in the Yup'ik saying "ak'a tamaani anguyiit anipaunguatullruut," which means "long ago warriors used to pretend to be owls." The Iñupiat show similar respect to Ookpik, around whom many teaching stories revolve.

These beings can be long-lived (the oldest known in the wild was a twenty-four-year-old female), but they aren't faring well. The North American population appears to be dropping precipitously, by as much as 50 percent in the past fifty years. What's clear—from depictions in European prehistoric cave paintings right up through Harry Potter's pet Snowy Owl, Hedwig—is that humans have long been enchanted by the regal Snowy Owl.

TOM SEXTON

Snowy Owl

I still see it rising from the marsh
no more than a foot from where I stood,

a rush of white, yellow eyes, cold flame
looking down. I stood motionless,

a hare caught out in the open,
and then it was gone, in the hollow

where it was resting, scat, a few
small pieces of bone, the rising cold.

Snow Mosquito

Culiseta alaskaensis

An old Alaskan joke is that our state bird is the mosquito (rather than the willow ptarmigan). In truth, Alaska's mosquitoes are both large and plentiful. According to one calculation, about seventeen trillion of them, collectively weighing some 48,000 tons, hatch in Alaska each year. Of the thirty-five species found in the state, the Snow Mosquito, who overwinters and is the first to emerge in spring, is the largest—and the slowest.

Most mosquito species overwinter as eggs in wet or moist places, where they go through stages from eggs to larvae to pupae before emerging as adults in spring. Alaska is a perfect place for mosquito breeding because of the statewide expanses of ponds, sloughs, and areas of standing water. The overwintering Snow Mosquitoes bury themselves as adults in leaf litter beneath the snow or in bark crevices and rotting trees, and—as the common name suggests—can emerge while there's still snow on the ground.

Mosquitoes, as well as many other Alaska insects, survive below-zero temperatures because of two biochemical processes. First, their body water is replaced by glycerol, which acts as an antifreeze and keeps their cells from freezing and rupturing. Second, their body temperature regulates downward in what's known as supercooling. The milder winter and spring temperatures that come with a warming climate are increasing mosquito survival.

Only female mosquitoes bite humans and other animals; they need the protein in blood to grow and nourish their eggs. The smaller males feed on flower nectar and berry juice. The female biters zero in on prey by following odor, body heat, and carbon dioxide from breath. As annoying as they can be, mosquitoes in Alaska do not carry diseases. One species that can transmit West Nile virus has been found in Alaska, but the disease itself hasn't—yet.

Mosquitos threaten caribou survival by forcing caribou to spend more time fleeing than eating. During migration, caribou are driven by mosquitoes to windy, elevated snow patches for relief. In spring, caribou migrate to windy Arctic coastal areas (such as the coastal plain of the Arctic National Wildlife

Refuge) for calving at least in part to protect newborn calves from mosquitoes. Mother caribou can be harassed enough to weaken, and young calves can die from the loss of blood from mosquito bites. Although a single mosquito bite withdraws only about one-fifth of a drop, they do add up. It's calculated that half a million mosquitoes could kill a naked human in three hours by sucking up a quarter of that person's blood.

What good are mosquitoes? Like every other member of the living world, mosquitoes play a critical ecological role, despite the harassment of their bites and their incessant buzzing, and despite the risks they pose for caribou. In their aquatic stages, they feed fish, wading birds, and other species. As adults, they're important pollinators as well as food sources for birds and bats. (Putting up swallow nest boxes and bat houses is a natural way of controlling mosquitoes around your own house.) Many birds choreograph their migrations, nesting, and chick rearing with the annual cycle of mosquito hatches. Snow Mosquitoes appear just as many bird species, depleted from their long migrations, arrive in Alaska for their short breeding seasons.

RAY BALL

You Say the Mosquitos Are Terrible This Year

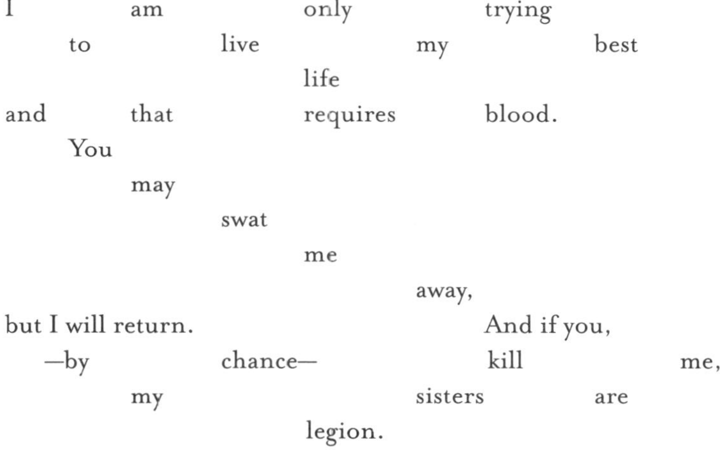

I am only trying
to live my best
life
and that requires blood.
You
may
swat
me
away,
but I will return. And if you,
—by chance— kill me,
my sisters are
legion.

Pingo

Rising from a treeless expanse of tundra, covered with mica-flecked soil that reflects sunlight, a particular Pingo's peak is tunneled with an old wolf den and pocked with ground squirrel holes like a high-rise apartment building. Trails radiate out in all directions from the Pingo's circular base.

Pingos, low hills or mounds with ice cores, are perhaps the most spectacular landforms associated with permafrost. Above this permanently frozen ground, soil and surface water form unique shapes. These unique landforms include raised tundra polygons from 10 to 100 feet in diameter, connected in vast honeycombed networks, and thermokarst phenomena appearing where permafrost thaws, creating sinkholes, tunnels, caverns, and ravines. Then there are Pingos.

Historically, about 80 percent of terrestrial Alaska has been underlaid by permafrost, with about 30 percent of this being continuous, 35 percent discontinuous, and the rest sporadic or isolated. It's in the continuous and discontinuous permafrost that Pingos are found. The process creating Pingos is much like frost heaving: groundwater freezes and expands, creating an upward pressure that causes the land to rise, creating the dome shape of a Pingo.

Pingos can be anywhere from 10 to 200 feet high and 50 to 1,500 feet wide. They grow slowly, taking decades or centuries to form. And they continue to grow as long as permafrost remains. Just south of Prudhoe Bay, Alaska's Kadleroshilik Pingo, the world's tallest, is 180 feet high and still rises by about an inch a year.

More than eleven thousand Pingos are known on Earth, of which Alaska has at least fifteen hundred. Where they exist in the boreal forest, Pingos are hard to spot. But where they tower over an otherwise flat tundra, they are conspicuous and make excellent lookouts and waypoints for people and wildlife. The summits of larger Pingos may contain crater lakes where their ice lens has partially melted. As tundra Pingos are often the only dry land around in summer, they're islands of diversity, providing shelter to an array of plants and animals, including shrubs, trees, birds, ground squirrels, foxes, and wolves.

However, as Arctic temperatures rapidly rise from global warming, permafrost temperatures are rising even faster—by 3 to 5 degrees F in the last forty years. As permafrost thaws, the unique terrain and ecosystems at the top of the world are being irreversibly changed. Permafrost thaw makes Pingos vulnerable to collapse, leaving behind remnant lakes. And as the ice layers beneath continue to melt, surface waters drain, ponds vanish, and shrubs and trees spread across what was once tundra.

Like small volcanoes rising above the vast Arctic tundra, these ice-core hills thrust from the surrounding terrain are iconic landmarks in permafrost, a world with irreplaceable richness in a thin layer of life.

CAROLYN KREMERS

Hungering to Discover All We Can

Pingo—*pingu*, from Greenlandic (Inuktun) and Inuit (Inuktitut), meaning *small hill*. First described by a European in 1851, in the form of a sketch by John Richardson in his two-volume book, *Arctic Searching Expedition*.

Out there to the west—
beyond the shining yellow leaves
drifting down, down, down—
a pingo has risen:

its icy core of water
pushed up
and frozen solid by permafrost,
its sides and dome-shaped top
likely insulated now

with black spruce skinny trees, twisted willows, alder cones,
squirrel middens, moose droppings, mushrooms deliquescing,
and the scarlet leaves of fireweed, now brown and dead.

I came upon this pingo in a scientific paper—
"Comparison of geophysical investigations for detection of massive
ground ice (pingo ice)," by Kenji Yoshikawa, et al. (2006)—
and I felt stunned.
It is possible that a large amount of water exists
in a solid phase beneath the surface of Mars,
and the geophysical methods presented here
may present useful tools for Martian exploration . . .

This pingo is my neighbor,
my own thousand-year-old neighbor!
Diameter 120 meters (394 feet), height 10 meters (33 feet) . . .
And it even has a name: the Cripple Creek pingo.

From my little house high up on Chena Ridge,
ringed with tall, wise birch trees,
of course I can't see the marvel down below.
Yet isn't it less than two miles away,
as the raven flies?

I've begun to hope
that my unusual neighbor can be located,
after freeze-up,
with a hand-drawn map—
and with my warm Sorels or skis
and my good friends, Chris and Tony.

One Saturday or Sunday
we will park
on a snowy gravel road
and set out cross-country, searching.

The ice-filled pingo will be ascending—
or collapsing—
on the valley floor.

Like life or a matchstick
it will be there,
rising up
from flatness and surprise,
as a *pingu* does.

And we—the curious,
the explorers, scientists,
tundra first peoples,
the missions to Mars—
we'll be hungering to discover
all we can.

North American Wolverine

Gulo gulo luscus

Compared to cousins mink, marten, and weasel, the Wolverine is bigger (up to 40 pounds), seldom seen, and little studied. Wolverines' reputation as fierce predators is mostly myth, although they do defend food sources and territories aggressively. They're primarily opportunistic eaters, scavengers who feed on kills made by others or the remains of winter-killed animals, but they also hunt for smaller mammals and birds and bird eggs. And they've been known to chase down exhausted caribou; on the North Slope, caribou are, in fact, a major food source.

Wolverines live throughout most of Alaska in extremely low densities. Their home ranges, depending on food availability, are often several hundred square miles. They're well built for snow country, with thick fur, short legs, wide bodies, broad heads, and large flat feet with retractable claws, making them excellent tree climbers. Their fur is a dark chocolate color with two cream-to-gold-colored stripes running from each shoulder to the base of the tail, and often a white patch on the chest and neck.

These beings have tremendous strength and endurance and can travel as much as 40 miles in a day, even in steep mountain terrain. Their strong sense of smell allows them to track kills for great distances, even under snow. With powerful jaws, large neck muscles, and inward-curving molars, Wolverines can crush bones and tear apart frozen meat. They can drag prey three times their own weight for considerable distances to cache food.

Remaining active all year, they're solitary except when raising kits. Wolverines most often den in snow caves, often including a complex of tunnels. Although mating occurs in the summer, embryo implantation doesn't take place until early winter and depends on the condition of the female. A well-fed female will birth two or three tiny kits (less than a pound each) in late winter. The kits develop rapidly and reach adult size by the following winter. Males visit dens during the nursing period; later, kits sometimes join their fathers to travel together. One male is often paired with several females, while other males have no mates.

The Wolverine's only natural predator is wolves, but humans also hunt and trap them to use their fur to trim hoods, both as ornamentation and for the guard hairs, which resist frost buildup. Starvation also is a leading cause of death. In the wild, individuals may live eight to ten years.

Alaska's Indigenous Peoples have long held the Wolverine in high regard, admiring this being's strength, endurance, cunning, and courage. Anthropologist Richard Nelson, in *Make Prayers to the Raven*, wrote that the Koyukon Athabaskans consider Wolverine "the greatest of all spirits, most demanding of obeisance and respect. It is a dangerous power." The Koyukon know the animal as Neltseel; the Iñupiat as Qavvik.

The Latin name *Gulo gulo* translates to "glutton glutton," which may refer to the Wolverine's style of voracious eating or may have come from a mistranslation between European languages, from an origin meaning "mountain cat." The common name is thought to come from "little wolf." Due to the strong and unpleasant odors from the Wolverine's anal scent glands, nicknames include "skunk bear" and "nasty cat."

The Wolverine population in Alaska is considered stable, except on the North Slope where it's in decline, possibly because caribou are declining; however, populations are difficult to estimate, given how spread out Wolverines are. In western Canada and the lower United States, Wolverines are a conservation concern, and they're exceedingly rare in Eurasia today. Their decline in most places is associated with the loss of wildlands—the expanses of open country and food sources on which they depend. Since Wolverines need snowpack for denning, climate change may be a new obstacle.

Wolverines are acutely wedded to life in the Far North; their primary requirement is wilderness. They're incredibly intelligent and resourceful and have been known to steal bait and animals from traps without getting caught themselves. They're also quiet, secretive, and wary—it's rare for us to see them, and when we do, these beings who epitomize wilderness disappear quickly.

SETH KANTNER

All About Being Gone

The snow, in a narrow valley north of the Amakomanak, is
soft and silent after the wind-bashed tundra; deep and marred
only by fading sinkholes made by a moose before the last
storm. The land is empty, hushed and hungry, white and

waiting with frosty lives locked beneath. Only the willow thickets reach their wooden limbs into the frozen world. The day is the grays and pinks and purples and bruised blues of winter, here miles and mountains beyond the last tree.

Ahead, the glowing snow is webbed with a fishnet of ptarmigan trails. The drifts are downy gauze, like feathers ready to ride the wind. Suddenly white wings do burst into flight, dizzying, as if the ground itself is moving. Fluttering white against blue, a flock of ptarmigan is up and away. And standing dark, robbed of a meal, eyes staring through you—motionless, for an instant—is the wolverine.

On the snow, the wolverine is black as a rip in the sky. Dark face already spinning, flashing teeth. Short legs, low drooping tail. Shy, this creature is all about being gone, a dark muscle designed to disappear. Already the willows have closed and reveal nothing.

Smaller, black on white the wolverine reappears, bounding tirelessly up the bluff, too vertical for most to climb. Especially you. It vanishes over the rise, invisible, crossing open ground as wolverines do.

Minutes later—wait!—a dark dot on the mountain. Climbing higher and higher, still bounding, ignoring gravity. Finally disappearing over a pass between peaks, gone for good. Left behind is a slurred trail, footprints each with five claws tearing aside whatever is in the way. Left behind is awe, for a creature who survives on old buried bleached bones, dried dead voles, sandy fish fins, and an occasional warm ptarmigan.

The wolverine is gone, beyond mountains, out there even now, hiding in folds in the land, traversing this huge hungry home, possessing a force, the endurance and strength of Other. And here, humbled, alone in silence and snow, you recognize determination. You thought you knew the word. You've just now witnessed the definition.

Alpine Azalea

Kalmia procumbens (synonym *Loiseleuria procumbens*)

First named *Azalea procumbens* by Linnaeus, this charming mountain plant blooms in early summer, setting alpine slopes ablaze with soft pink crown-shaped flowers. Alpine Azalea is a tough little plant, as alpine plants must be, to grow in rocky, dry soils on wind-whipped slopes. Like dwarf willow and dwarf birch, this plant spreads woody limbs across the ground, branching and curving just like an erect tree but always prostrate (hence *procumbens*), never taller than a couple of inches.

This being is usually found in the company of lichens, especially among the pale-green froth of reindeer lichen. A member of the heath family, of whom there are many in Alaska, Alpine Azalea is in the same genus as mountain laurel but is the only member of this genus who doesn't have spring-loaded anthers that are triggered to catapult pollen when stepped on by an insect.

Alpine Azalea does have unique adaptations to the wind-scoured mountaintops where the protective blanket of snow is often blown away, leaving plants to freeze and dry out. This azalea holds onto tough, leathery leaves all year and grows in tight mats to retain moisture and heat and protect small buds in spring. A surprisingly wide rootstock and long taproot anchor deep, putting most of the plant's weight underground and allowing a single plant to live up to sixty years.

Alpine Azalea is not tough enough, however, to withstand trampling, and is extremely vulnerable to climate change. Like all plants in the heath family, Alpine Azalea has a symbiotic relationship with fungi and relies on insect pollinators. But if no pollinators appear with pollen from another plant, the stamens bend against the stigmas and self-fertilize—almost like a catapult.

SUZI GOLODOFF

Haiku

in the fog-drenched wind
reaching out sideways to bloom
alpine azalea

Willow Ptarmigan

Lagopus lagopus

Like the snowshoe hare and the arctic fox, the Willow Ptarmigan (pronounced TAR-mi-gan) is a master of disguise—snowy white in winter and brown in summer. In their winter plumage, males and females look identical—at least to humans. Beginning in early May, males develop capes of chestnut-red feathers with which to impress mates; they also sport bright red eye combs. Females change more fully from white winter dress to a mottled brown that blends well into backgrounds. A distinctive characteristic of both sexes is their feathered

feet, which help keep them warm on snow; they also use their feet to excavate tunnels in snow, in which they take shelter from storms and bitter cold.

Of the three ptarmigan species found in Alaska, Willow is the largest and most numerous. (The others are rock ptarmigan and white-tailed ptarmigan.) The Willow Ptarmigan was chosen to represent Alaska by schoolchildren in 1955 and became Alaska's official state bird in 1960. The Willow's range across the north includes Canada, Scandinavia, and Russia; this being lives in most of Alaska except for areas of thick forest and the far Aleutian Islands and prefers alpine and arctic tundra.

"Willow" is an appropriate name for these ptarmigans. They nest under willows and depend on them for food. They eat willow leaves in summer and buds, twigs, and catkins in winter. Berries and various seeds are other important foods.

Females lay six to ten eggs, which hatch in late June or early July. In a departure from male roles in every other ptarmigan species in the world, males share parenting duties by helping to defend the broods. Females typically try to draw predators away by feigning injuries, and males dive aggressively at intruders.

These birds are highly social. In winter, they can form flocks of several thousand where food is plentiful. When in groups, they play together—extending and bobbing their heads, flapping their wings, and jumping around erratically. They migrate between ranges seasonally, sometimes by just a few miles up or downslope and sometimes by more than a hundred miles. They are not shy and often strut along road edges, where they're at risk of being hit by vehicles.

The Willow Ptarmigan's low-pitched voice can sound like a human, with a large repertoire of clucks, cackles, chuckles, growls, and peeps. Males, when impressing females with tail fanning, wing spreading, and bowing, also make barking and rattling noises. The Yup'ik know this bird as Aqesgiq, which replicates the male's mating call.

The bird's name comes from the Scottish Gaelic word *tarmachan*; the *p* was added because early ornithologists mistakenly thought that the word came from the Greek *pteron*, meaning "wing." *Lagopus lagopus* means "hare-footed" in Greek, in reference to the bird's heavily feathered feet and toes. In Alaska, ptarmigan are sometimes called tundra chickens, and the story (or perhaps myth) goes that the gold rush town of Chicken in eastern Alaska was named in 1902 after the ptarmigans who were both common and an important food source but whose name was thought too hard to spell.

The Willow Ptarmigan's winter plumage works to camouflage this being only when snowfall arrives on time and remains. Inconsistent and reduced snowfall due to climate change is now putting Alaska's state bird more at risk from predators. The northward movement of forests also affects their habitat.

In late summer, when Willow Ptarmigan chicks have fledged but the family remains together, it's common for hikers to be surprised by a sudden burst of sound and wings from nearby willow thickets as the entire family flushes, flies in a low curve for a short distance, and then lands, disappearing once more into the landscape.

SUSHEILA KHERA

Refuge in Illusion

Plump lumps of snow settle
on the willow branches
at dusk. A sudden fluttering
and off they go. Some things
evade us in plain sight.

Buds and seeds, gathered remnants
of the autumn past, fill the craw.
A little sachet of food resting
above the heart. Some things
are held close and warm.

A hoarse cackling skitters
through the air. Winter coloration
quietly turns to a speckling of browns,
blacks, hints of red, spread over head,
neck, back, the hare-like feet.

A flock flies, not too high, glides,
disappears in low shrubs on thick moss.
Through how many millennia can
a disguise offer safety when on short notice
the pace of the world changes?

To a bird standing still, exposed
on the monochromatic band
of summer road
the dappled light
might seem a shelter.

Arctic Ground Squirrel

Spermophilus parryii

Every fall, across Alaska's alpine and Arctic tundra, Arctic Ground Squirrels retreat into their burrows just a few feet belowground. They curl up in their nests of lichen, grass, and bits of fur from caribou, mountain goat, or Dall sheep, and they hibernate—for up to nine months in the more northerly part of their range. Without food or water, under outdoor air temperatures that can dip to almost –80 degrees F and howling winter winds, these 1-to-3-pound mammals annually perform this amazing biological feat, an extreme adaptation to harsh northern weather.

Before hibernation, Arctic Ground Squirrels somehow flush particles from their tissues that might serve as nuclei for ice formation. In their underground hibernaculum, these beings conserve energy by entering into states of torpor in which their metabolic rates and body temperatures drop drastically—their body temperatures falling from 99 degrees to 27 degrees F—for up to three weeks at a time. They are the only vertebrate known to survive such sustained low body temperatures. Their oxygen consumption can also drop by 90 percent, and their hearts can slow to three beats per minute. Throughout hibernation, they periodically arouse from torpor for a couple of days at a time to use their stored fat, shivering to generate heat and bringing their body temperatures back up to normal. By spring, both males and females have lost about a third of their prehibernation weight.

Arctic Ground Squirrels live in colonies across most of Alaska, primarily in dry, open alpine and Arctic tundra at elevations that range from sea level in the Far North to high in Southcentral's coastal mountains—anywhere with loose soils that provide early vegetation and allow for burrow construction. These round-headed mammals are built for burrowing and digging, with strong forearms and legs, sharp claws that can be longer than their paws, and soft pads on their handlike front paws for manipulating both dirt and food.

Males (called boars) aggressively defend territories where multiple females (sows) reside. After breeding season, boars suffer significant mortality from weight loss and stress, and sows group together in kin clusters. Litters, born in May, vary between two and ten kits; the kits grow rapidly and leave their dens after eight or ten weeks.

Arctic Ground Squirrels are generalist feeders, with a diet that includes not just plant stems, leaves, roots, fruits, seeds, grasses, sedges, and mushrooms,

but also bird eggs, chicks, invertebrates, carrion, and even small mammals like lemmings. They cache food in burrows to eat when they rouse from hibernation.

As a keystone species, the Arctic Ground Squirrel is an important food source for many predators, including wolves, brown bears, foxes, and owls, which especially turn to them during low cyclic phases of other prey. Their burrowing increases soil looseness and its organic content, nutrient levels, and seed germination rate, which in turn increase plant productivity. Their feces and urine also fertilize the soil and support plant growth.

Of the forty ground squirrel species in North America, the Arctic Ground Squirrel is the largest and the only one found in Alaska. This squirrel is also known as a Parka (or Parky) Squirrel because the fur, brown or tan with white spots, is used in traditional clothing, including women's fancy parkas. The name in Iñupiaq is Siksrik, after the chattering alarm call.

When out on the tundra, listen for the whistles and calls of Arctic Ground Squirrels. These beings are hypervigilant and frequently sit or stand on their hind legs to listen and look around them. Normally, one of them on the lookout will warn others of possible predators. A high-pitched whistle warns of aerial predators like owls and hawks; a guttural chatter identifies a terrestrial one, like us two-footeds.

BILL SHERWONIT

A Most Unusual Encounter

Over the years, I've had several unusual experiences while exploring McHugh Peak, a favorite mountain of mine near Anchorage. Among the most curious of those was a midsummer encounter with an arctic ground squirrel.

I'd reached the top of McHugh—a mix of jagged outcrops, rubbly rocks, and ground-hugging plants—and was descending its northern spine when I happened upon the squirrel, hunched on the high alpine tundra within inches of the trail.

The squirrel's presence at first stunned and then delighted me, because of both his size and behavior. No bigger than a chipmunk (though plumper), the ground squirrel was by far the smallest I'd encountered in my decades of wandering Alaska's wildlands. Instead of dashing for his den as ground squirrels normally do, he remained perfectly still while I approached to within inches of his tiny body, then squatted for a closer look and even took a few pictures.

Even stranger, I could see no evidence of a burrow anywhere in the area.

I later learned that Alaska's ground squirrels are usually born in late May or early June. Eight to ten weeks later—between late July and early August—the young disperse from their dens. This ground squirrel and I met on July 6.

I suppose his parents could have mated early; or perhaps he was an exceptionally precocious pup.

That still doesn't explain the squirrel's behavior. Was he "frozen" by fear? Or hoping that I would not notice him, despite my close approach? The squirrel appeared to be in good health, so I wished him well and resumed my descent, carrying one more mystery down the mountain.

Moss Campion

Silene acaulis

Small plants with long lives, Moss Campions thrive in the most inhospitable of places. These circumpolar tundra and alpine plants survive subzero winters, heavy snow, high winds, short growing seasons, and rocky, dry, nutrient-poor soils. Long, thick taproots burrow between rocks and reach deep underground to find moisture, and a tight, mosslike shape retains heat and water.

One of the "cushion plants," named for their growth habit, Moss Campions make tidy ground-hugging mounds with only small bright-green leaves exposed to weather. Their ability to retain both the warmth from the ground and radiant heat from the sun creates temperatures within the cushion as much as 25 degrees F warmer than the surrounding air. Moisture and temperature moderation protect tender parts like flower buds and provide refuge for other small beings: their tussock shape can cut wind speed by up to 98 percent and hold moisture at levels up to 90 percent higher than the surrounding air. What's more, dead petals and leaves fall inward and compost to increase soil nutrients in their small circle of the world. Because they alter the environment to the benefit of other beings in their ecosystem, the Moss Campion is considered a foundation species.

While they're easily overlooked most of the year, in early summer, when the snow leaves and sunlight intensifies, tiny buds nestled under leaves burst forth, transforming indistinct mounds into brilliant bouquets of fuchsia-colored flowers. These half-inch blossoms first open on the south side of the cushion, providing their other common name: Compass Plants.

As with many beings at high latitude and altitude, Moss Campions are facing a dire future from climate change. And the harsh conditions where they live make it extremely difficult for new plants to establish. Fortunately, these tiny beings live for hundreds, if not thousands, of years: in Alaska, many reach 300 years of age, and the oldest here is estimated to be more than 350 years old.

So watch carefully for small, unassuming mounds of moss-like leaves. Tread gently around them, for they could well be older than your great-grandparents. And if you're lucky enough to see their blooming extravagance on some high rocky ridge, kneel down and inhale their lilac-like scent, even as you notice, too, the bees and butterflies drawn to their rare nectar.

CHRISTINE BYL

Campion

Moss campion grows above tree line, not down here in the taiga, where, in early May, along the path to the outhouse, your gravestone is still submerged in late-spring snow. Three feet beneath tundra, your white and gray fur has gone to gray silt, white ice. A sled dog litter named for wildflowers—your brothers *Sorrel, Larkspur, Lupine,* stalky blooms—and you got *Campion*, a lilt, a tease. Such contrasting creatures—you, the dog who could bear a heavy sled uphill, legs as long and strong as pistons, and moss campion, that tidy ground-hugger, a pale spray of stars pinned in a cushion of green.

Tough dog, stoic, but beneath your muscled coat you could be slow and tender. Asleep before a woodstove in winter, fur hot to touch. Tongue, teeth, so delicate, when taking treats. Softened by age and mischief, gleam in your glow-blue eyes: we laughed and laughed.

Your flower, the opposite—moss campion far tougher than it looks: pastel corollas, sturdy and undominated. Rosy alpine overwinterer makes its own heat beneath snow, while I nestle under layers and layers and must return, always, to warm.

Campion, I picture you still in summer above tree line, sprinting over tundra, then slowed by steepening grade. You moved paw by paw through gray and white granite boulders and then, while we drank water at the pass in the sun, you rolled and lolled atop your mossy namesake like a husky drag queen, the delicate cape cast under your shoulders, bright pink, a flounce, a flare.

Whimbrel

Numenius phaeopus

The elegant Whimbrel is one of five large sandpipers who breed in Alaska's Arctic and subarctic. Rarely observed in their remote nesting habitat, Whimbrels are more often spotted along beaches and tidal flats during flock migration to and from wintering grounds. Whimbrels are superb flyers and have been known to fly 2,500 miles nonstop over water or 5,400 miles in a single migration.

Whimbrels are distinguished by their striped heads, their loud *pi-pi-pi* whistling, and their very long, downcurved bills. *Numenius* means "new moon" in Greek and references the crescent-moon shape of the bill. *Phaeopus* means "gray-foot" and describes the color of the Whimbrel's legs. The common name originated in England, apparently as an interpretation of this being's call. Whimbrel

has two Yup'ik names—Kikikiaq and Pipipiaq, both of which sound like their flight call. Whimbrel's Iñupiaq name, Sigoktuvak, translates as "big, long-billed shorebird."

While wintering along coastlines as far south as Chile, and while in migration, Whimbrels feed largely on intertidal crabs and other marine invertebrates, using their fabulous bills to reach into sand and mud. On their Alaska summer breeding grounds, both parents and chicks feed mainly on berries and insects.

In Alaska, Whimbrels perform showy courtship behaviors, involving circling, chasing, dramatic aerial flights with sharp ascents, and gliding accompanied by a whistling and bubbling song. They nest on open tundra and in wetlands, usually on ridges or mounds of vegetation and often sheltered from the wind with small bushes. They lay three or four eggs, which hatch within a month, and the downy chicks, with their tan-brown mottled coloring keeping them camouflaged, leave the nest almost immediately.

Because they nest in remote areas, not much is known about the Whimbrel's population status, considered for now as of low conservation concern. Threats include environmental contamination and the loss of winter and migration habitat due to sea-level rise, erosion, and coastal development.

During migration, Whimbrels join thousands of other shorebirds of different species, towering over the more numerous but much smaller dunlins and western sandpipers, all of them feeding and resting together.

MIKE BURWELL

Curry Ridge

The lakes are without names
because they are high in rock,
fed only from melting snow.

Their shores fill
with repeated notes
of plovers.

The silhouettes of whimbrels
mimic the curve of the planet
with beaks as long as your hand.

Our eyes need to play
on such water and rock.
We say we will come back.

Sastrugi

In the snow-covered Arctic, as well as in Antarctica, the snow's hard-packed surface is shaped by the wind into features known as Sastrugi. Aside from being sculpturally beautiful and often tricky to negotiate on skis or sleds, these wind-eroded snow surfaces can be lifesaving for those who know how to read them.

After a storm, Sastrugi provide a guiding compass—a record of wind direction—especially in poor visibility.

Sastrugi (sometimes *Zastugi*) is the plural of the Russian word *Sastruga*, which translates to "small ridge." In appearance, Sastrugi can be compared to sand dunes or frozen waves on a frozen river, although their formation processes are not the same. The sides of Sastrugi facing the wind are steep and rise to points; the sheltered sides slope gradually. The intersection of curved surfaces can result in sharp edges, scallops, or shapes like anvil heads leaning into the wind. Sastrugi can become quite hard, almost like ice, through sublimation (solid becoming vapor) and recrystallization. They can range in height from just a few inches to several feet.

Wayfinding based on Sastrugi and many other environmental cues continues to be, even in our modern GPS age, a skill of Alaska's Indigenous Peoples. Today's backcountry recreationalists also use Sastrugi to identify the leeward side of slopes, which can be prone to avalanches. And, while it's not known for certain, biologists suspect that wild beings like polar bears also read Sastrugi to find their way.

TODD SFORMO

Mind of Winter

The crisp, creasing lines of sharp-angled sastrugi stretch only a short distance in front, because ankle-high blowing snow smokes by. Together they bewilder the snow into flickering flame—without the crackle of burning wood. Under these conditions, snow particles bump into each other, displace one another, hopscotch in long lines out of sight in a movement called saltation. Saltatory propagation also defines the movement of human electrical signals, an action potential jumping from one node to the next down myelin nerves. The mind's gazing at this wordy connection binds winter to my cognition, as the blizzarding lines propagate thoughts away from me as forgetting.

Arctic Fox

Vulpes lagopus

Arctic Foxes, the Arctic's smallest carnivores, are inhabitants of cold, windswept open spaces. They thrive in treeless expanses throughout the Arctic, as well as along Alaska's western coast. In winter, they range far out onto pack ice, following polar bears to scavenge scraps from their kills.

Weighing only 6 to 10 pounds, these tiny, agile, cat-size beings are smaller than their red fox relatives. That they can survive the Arctic winters without hibernating is thanks in large part to their compact bodies, short ears and muzzle, and dense winter fur. Even their paws are covered in thick fur in winter—hence the Latin name *lagopus*, meaning "rabbit-footed." An incredibly keen sense of smell and acute hearing allow Arctic Foxes to find food deep under snow, also contributing to their ability to endure bitter Arctic winters without hibernating.

Arctic Foxes occur in two color phases—white and blue, with white being most common on the mainland and blue on the Aleutian and Pribilof Islands, although young of both phases may occur in the same litter. (It's thought that color phases are related to the overall color of the environment, serving as camouflage.) Blue-phase foxes remain bluish, brownish, or charcoal-colored year-round. White-phase Arctic Foxes shed their long, thick white winter fur in early April to wear short, thinner brown summer fur, and then switch back to pure white in time for winter's first snowfall.

Arctic Foxes mate for life. Their dens, often the enlarged burrows of ground squirrels, are dug into sandy, well-drained soils of low mounds and river cutbanks, or sometimes under rock or driftwood piles. In these dens, which can extend for up to 12 feet underground and have multiple entrances, litters of up to seven pups are born in May; both parents feed and rear the young. These canids, small as they are, may travel in family groups for long distances, as much as hundreds of miles seaward in fall and inland in early spring. They communicate with one another with high-pitched barks and are also known to purr like cats. Arctic Fox lives are not easy, and although they can live for ten years, only one in twenty-five normally survives past age four.

Arctic Foxes can leap several feet into the air before pouncing on their prey. They're omnivorous, feeding primarily on lemmings and voles in summer, but also on nesting birds and their eggs, berries, and the scavenged remains of other animals. When food is plentiful, they store it in caches for later use. As a top

predator in the Arctic, they thrive when lemming populations are high in their cycle. They play another important ecological role in distributing nutrients across the Arctic for plant growth.

Alaska's Indigenous Peoples consider the Arctic Fox (Tiġiganniaq, "the little white one" in Iñupiaq) to be a helpful neighbor who leads humans to food hidden beneath snow and announces the approach of polar bears. The lightness of this creature's movements has been linked in Indigenous beliefs to escorting souls to the swirling dance of the aurora borealis.

Arctic Foxes, especially of the blue phase, were introduced to many of Alaska's islands by those eager to cash in on their pelts, first by Russians starting in about 1750 and continuing during the fox farming era of Alaska's history, from the 1890s to the 1930s. More than 450 islands were involved in what was for some time a lucrative fur business, to the detriment of ground-nesting birds and their eggs. In recent years, some of these introduced populations of blue-phase Arctic Foxes have been eradicated from some islands to allow native bird populations to recover.

As the climate warms and shrubs and trees move northward, red foxes are moving with them. Both larger and more aggressive than Arctic Foxes, they've been invading dens to prey on Arctic Fox pups and adults and take over their dens. Other threats to Arctic Foxes include mining and oil and gas development.

Camouflaged against winter's snow, his small white body curled into a ball and his fluff of tail curled over his petite triangle of a face like a blanket, the tiny Arctic Fox endures the deep cold, looking for all the world like a mound of snow himself.

TIFFANY ROSAMOND CREED

Moonlit Children

In April and May, we bloomed:
You, who holds a whole hour in your palm
as if to coax it like a frightened creature,
and me, leafing through a year like paper.
Lemmings fell from the sky like rain,
you swore it up and down and so it was.

Still, I struggle to pin the sun down,
wooly-minded and cold, but okay with it,
scarcity a scarf wrapped around my neck.
Some arctic foxes do not survive beyond their first year.
Some make it to eleven. "Not enough time," you said.

We are moonlit children, the choir of the biome.
We share the same memory and then some.
Do they remember the velvet plateau?
Does *not enough* cross their minds for very long?
Their rounded bodies curled in their hollows:
Summer morph and all, winter morph and all,
the agency of ice, and how it carved the earth.

Lingonberry

Vaccinium vitis-idaea

In Alaska's autumn, berries seem to be everywhere. At least fifty different berry-forming species grow in every kind of terrestrial habitat. There's the plump orange salmonberry shrub reaching high in the coastal rainforests, the northern red currant (also called gooseberry) who thrives in moist woods throughout most of the state, and kinnikinnick, a sprawling recumbent evergreen of dry woods and open areas, just to name a few. And there's Lingonberry, who ranges throughout the Arctic and subarctic around the globe and into some of the northern Lower 48.

Ethnobotanists say that the more names a plant has, the greater the cultural importance. Lingonberry goes by many names—Indigenous names and English names that depend on heritage and geography. These include Kikmiññaq (or Kikmiñat or Kimmigñaq) in Iñupiaq, Hey gek'a ("winter berry") in Dena'ina, and Kenegtaq in Alutiiq. English names include Low Bush Cranberry, Rock Cranberry, Mountain Cranberry, Partridgeberry, Cowberry, and Foxberry. The names hint at the plant's habitat and relationships.

A member of the heath family, Lingonberry is a small evergreen shrub who grows low to the ground, usually no more than a foot high. The plants grow in a clonal colony, and individuals begin producing flowers and berries when they're about five years old. Lingonberry thrives in the acidic soils of bogs and open woods as well as in tundra, where, in Alaska, blueberries and crowberries are companions.

The pale pink blossoms are shaped like a bell. The berries, which ripen in September, are red in color and tart in taste, getting sweeter after a frost. The bright berries remain on their stems throughout the year. They're an important winter food for dozens of species, from polar bears to ptarmigan to pikas; in some parts of Alaska, they're a critically important food for moose. Based on their nutritional profile and antioxidant content, they're known by humans as a superfruit.

Lingonberries are also important traditional food and medicine for Indigenous Peoples. Dena'ina scholar Peter Kalifornsky reported that "the old people said that there is more nutrition in the low bush cranberry than in any other berry." To offset their puckering tartness, these berries are most often used in a sweetened sauce, baked goods, preserves, and liqueurs. Akutaq (a Yup'ik word meaning "mix it together" and sometimes called Eskimo ice cream) combines Lingonberries and other berries with seal or caribou fat, sugar, and sometimes bits of meat or fish. The berries store well year-round when frozen or kept in a cool place, or dried.

With their cheerful clusters of brilliant red berries nestled among oval-shaped waxy green leaves, Lingonberry plants brighten Alaska's bogs, forests, and alpine areas year-round.

MARIE TOZIER

Lingonberries

Creep above
Autumnal tundra

Bright red kernels
Tart on the tongue—

When fully ripe
Of burnt claret.

Musk Ox

Ovibos moschatus

What looks from a distance like dark boulders in a sea of green becomes at closer range a scene right out of prehistory: shaggy mammals grazing, moving slowly, muzzles deep in the thick summer meadows of Alaska's undulating tundra. These are Musk Oxen, rugged survivors of the last ice age.

They're also resilient survivors of some of the harshest weather on Earth. Unlike almost all other Arctic animals, who hibernate or migrate in winter, Musk Oxen endure months of darkness, howling blizzards of blowing snow, and temperatures below –40 degrees F in the open, unsheltered tundra. While in summer they feast in wet meadows, in winter they climb higher to ridgelines where wind blows away heavy snows, allowing them to dig down for lichen and other vegetation.

On these high ridges, however, they also experience increased wind chill. Musk Oxen save energy by standing or lying still; they reduce glare from sun reflected off snow and ice with their horizontal slit-like pupils; and they slow their digestion, breath, and heart rate. They become mounds of fur covered by snowdrift, waiting out winter for spring's thaw.

Perhaps their most important adaptation for braving winter is their fur. Coarse outer guard hairs hang like curtains almost to the ground, shedding water and keeping insects at bay; these guard hairs also protect softer inner hairs, called qiviut. Qiviut is considered the rarest fiber in the world: not only is it one of the warmest furs, but it's also stronger and more durable than sheeps' wool and softer than cashmere. Musk Oxen shed qiviut in summer, and tufts stuck to willow branches or grasses are collected by crafters to make hats, scarves, and mittens. A cooperative of Alaskan Indigenous knitters produces garments from qiviut gathered on the Seward Peninsula and in western Alaska, with each village having its own signature pattern.

Referring to their long guard hair, Iñupiat call the animals Umiŋmak (also spelled Oomingmak), meaning "the animal with skin like a beard." The English common name refers to the distinctive odor males produce to attract females during mating season.

Stocky and compact, with cloven hooves, Musk Oxen are the world's largest goats. Mature bulls reach about 5 feet in height and weigh up to 800 pounds, while females are slightly smaller. Both bulls and cows have wide, thick horns that meet in the middle of their heads and curve around their faces like a helmet, then curve out to sharp points. Males head-butt and tangle horns during the fall mating season, bellowing like lions and crashing together with resounding force. They're protected from the shock of impact by 3-inch-thick skulls covered by 4-inch-thick horn bosses (as the large base of the horns is called).

These horns, with their sharp protruding tips, are also their defense system. Living in complex social groups of up to seventy-five individuals, Musk Oxen when threatened will form a circle, with their horn-bearing heads facing out and lowered, and their calves in the center of the wheel. Against their natural predators, from prehistoric animals to today's wolves and bears, this defense works well. Against humans with rifles, it does not.

Whalers, explorers, and other nineteenth-century travelers to Alaska's Arctic found Musk Oxen easy prey and often killed an entire herd where they stood in their defense circle. This large-scale killing wreaked havoc on their populations. By 1900, the once-plentiful Musk Oxen had disappeared from Arctic Alaska.

At that time, Musk Oxen remained only in Greenland and Arctic Canada, sparking concern for their extinction. So in 1930, thirty-four calves were relocated from Greenland to Fairbanks and then a few years later to Nunivak Island. From that initial Nunivak herd, they were released to mainland Alaska. Now more than five thousand Musk Oxen range throughout most of the treeless Arctic, from the Yukon-Kuskokwim Delta to Kaktovik.

In recent years, however, herds in the Arctic Refuge and nearby areas have declined; scientists suspect this may be due to warmer winters causing ice layers and deeper, heavier snow, keeping Musk Oxen from being able to dig for food. As well, changes in vegetation and the northward movement of tree line are altering their habitat.

Musk Oxen once roamed the northern plains with saber-toothed tigers, dire wolves, and woolly mammoths, standing their ground against massive predators. Living relics from a prehistoric age, they alone survive, reminders of the power of resilience and of the deep history of this planet we call home.

BARBARA HOOD

Return

As far from people as we have ever been,
as far to the horizon as we have ever seen,
the Arctic river flows heavy with silt, slow
and rasping against our hull, our paddles,
a soothing harmony of quiet and carving—

this channel, this valley, these mountains
that rise worn and wise, like grandparents,
as if the landscape agrees to be taken
in the wide web of rivulets gathering
grain by grain away to the distant sea.

The ancient ones left black chert, far
from its origins, high on a hunter's ledge,
left fire rings and house pits, left names,
left legends—pressed their feet into this
riverbank, their hands into these hills.

We sense no consent in their leavings, only
echoes of longing: a lone musk ox climbs a
ridge and spooks a herd of sheep, returned
after a century a stranger to his kin, stepping
sure-hooved to the place where he belongs.

Alpine Forget-Me-Not

Myosotis alpestris

With five sky-blue petals offset by bright white-and-yellow centers, Alaska's charming state flower grows high in the mountains between 7,500 and 10,000 feet, tucked into subalpine and alpine meadows. At 5 to 12 inches tall, this plant displays a true blue color that's rare in the world of flowers.

Alpine Forget-Me-Not is one of more than 150 different species of forget-me-not around the world with the same common name. This name is the English translation of the French *ne m'oubliez*, and these plants are associated worldwide with remembrance. The seedpods of this perennial in the borage family hitch a ride on a passing mammal's fur or human's clothing to ensure wide dispersal. Several other species grow in Alaska, including mountain forget-me-not and splendid forget-me-not. Another, *Myosotis sylvatica*, easily adapts to gardens and yards, merrily spreading pale blue, pink, and white flowers beyond all boundaries. Alaska's state flower is not so easily propagated. Like many alpine plants, the Alpine Forget-Me-Not prefers the dry, thinly nourished soils of high mountains to the rich, dark humus of garden soils, where other plants quickly outcompete them.

While the flowers are stunning, it was the delicate plant's hardiness and resilience that inspired Alaskans to make Alpine Forget-Me-Not the state flower. Even before Alaska's 1959 statehood, at the turn of the century the plant was adopted as an emblem of a pioneers' organization. Then in 1917, a few years after Alaska became a territory, the Territorial Legislature approved the Alpine Forget-Me-Not as the official floral emblem of the Alaska Territory.

This alpine wonder is also represented in the Alaska state flag. In 1926, when territorial governor George Parks was lobbying for Alaska to become a state, he realized other states had their own flags, so he convinced the Alaska American Legion to host a flag design contest for Alaska students from grades seven to twelve. The winner was Benny Benson, a thirteen-year-old Alutiiq boy in Seward. Of the flag's design, Benny wrote: "The blue field is for the Alaska sky and the forget-me-not, an Alaskan flower. The North Star is for the future state of Alaska, the most northerly in the union. The Dipper is for the Great Bear—symbolizing strength."

This tiny, bright flower is embedded in Alaska in other ways as well. Alpine Forget-Me-Not is often portrayed in Indigenous beadwork, sewn into tunics, regalia, parkas, mittens, and jewelry. Blue as the sky, resilient and strong enough to withstand life in Alaska's wind-and-snow-scoured mountains, Alpine Forget-Me-Not is certainly a plant worth remembering.

KATHLEEN TARR

He Kept Returning

for Shawn Lyons, in memoriam

I have dreams of mountains,
and of a mountain man
I once knew.
On rain-soaked days,
he climbed the peaks,
and trekked through
canyons and ridges upon
unmarked and muddied trails.

In the silent awe
of the front range,
too many times to count,
he scrambled, often alone,
into high mist
and scattered snowfields
few have known.

Over decades of Junes,
in shorts and a windbreaker,
with an age-old rucksack,
he kept returning.

Hours of hiking
and bouldering,
he sat down on stone
to fuel his thin body,
with only hard candies.

What did Alaska's grand
mountains tell him,
what spell cast by talus and tarn,
and miles of summits reached?

And what of dainty
forget-me-nots
hugging rocks,
their tiny blue petals
tucked away
on sub-alpine slopes?
Did those small flowers,
delicate, enduring,
charm him, too?

White-crowned Sparrow

Zonotrichia leucophrys

The White-crowned Sparrow, one of twelve sparrow species who migrate to breed in Alaska, is ubiquitous—though more often heard than seen—throughout the state in spring and summer. The adult White-crowned is distinguished from other sparrows by a beautifully crisp black and white head stripe. The song, while showing incredible diversity, has a somewhat lilting melody of descending notes.

These birds favor brushy habitat, where they skip and hop through woody tangles or forage on bare or grassy ground nearby, feeding on seeds and, in summer, insects and spiders. When foraging, the birds are known for their unique habit of "double-scratching"—hopping backward to turn over leaves and then forward to pounce on what they uncover.

They nest in low bushes in alpine meadows and along forest edges; on tundra they nest on the ground, hiding their nests next to clumps of grasses or other plants. The females build cup-shaped nests of twigs, grasses, and mosses and lay three to seven eggs. In Alaska's short summers, they typically raise just one brood, although they may raise two or even three when conditions are ideal. Both parents feed the nestlings with insects.

After the breeding season, these Alaskan songbirds travel south to winter in southern California and Mexico. One was tracked flying an astonishing 300 miles in a single night. Pairs separate during the winter, with females generally traveling farther south than the males, but most pairs reunite back in Alaska for breeding in their homelands.

The remarkable songs of the White-crowned Sparrow have been the subject of considerable study, and biologists have learned that these birds actually have site-specific dialects. The young birds learn the basics of their songs in the first two or three months of life—not just from their parents but also from the entire song environment of their birthplaces. As salmon imprint on natal streams through smell, White-crowned Sparrows appear to imprint on natal nest areas through sound. Since adults return to their same home areas for breeding, they continue singing in their same dialects, although the songs in a region change over time—similar to what happens when humans play the game of "telephone." White-crowned Sparrows who live at the borders between two dialects may be bilingual, able to sing in both dialects. What's more, research indicates these little birds can remember previous songs: a 2020 study of White-crowneds in the San Francisco Bay Area found that during the quiet of the pandemic, the birds quickly rearranged their songs into their original preferred frequency, one they'd previously abandoned because of traffic noise.

Another area of study is the birds' unihemispheric slow-wave sleep, which allows them to stay half awake for up to two weeks while they migrate. (That is, one half of the brain sleeps while the other half remains alert.) It's hoped that understanding this astonishing sleep mechanism, which is shared by a number of other beings including dolphins and seals, might be helpful to humans.

It's familiar and welcome, the song of the returning White-crowned Sparrow, pouring out clear and sweet like streams released from winter's freeze, mixing in with all the other returning and year-round songbirds—music with a complexity that we humans are just beginning to understand.

AMY CRAWFORD PUREVSUREN

After a Storm

In the arctic pre-dawn,
the valley drained of song,

gaunt spruce boughs shiver
in the final gusts

when a sweet brave whistle
catches clearly in the ear

and bursting from willows
that lace the cobbled creek,

white crowns rise and scatter,
a chorus of wings flashing

in low golden rays, igniting
grass-tufted tundra,

limestone ledges chiming
with trilling sparrow song,

the morning wide-eyed
and so full of heart.

Gratitude

We thank, first of all, the place we know as Alaska and the magnificent large and small beings with whom we share it. Major thanks go to the Alaska poets, writers, and artists who contributed to the work. Along with us, they are deeply connected to their home places, and their talents and sensibilities make the book the beauty that it is. We very much appreciate their taking the journey with us.

We thank as well the many, many individuals and organizations that are the sources of Alaska natural history and cultural material. Some of those sources are listed in our Resources section.

At the start, we appreciated the support and advice of Elizabeth Bradfield, CMarie Fuhrman, and Derek Sheffield, co-editors of *Cascadia Field Guide: Art, Ecology, Poetry*.

Early on, Emily Wall, Rico Lanáat' Worl, Ernestine Hayes, Vivian Faith Prescott, Peggy Shumaker, Erin Hollowell, Julie Decker, Asia Freeman, and Marie Tozier all shared valuable suggestions.

Our appreciation extends to two other anthologies that led the way: *The Sonoran Desert: A Literary Field Guide,* edited by Eric Magrane and Christopher Cokinos, illustrated by Paul Mirocha, and *A Literary Field Guide to Southern Appalachia*, edited by Rose McLarney and Laura-Gray Street.

We're very grateful to Rick Steiner, Vivian Faith Prescott, Marilyn Sigman, and Richard Carstensen for their time and expertise in providing discerning and essential fact-checking.

Kate Rogers was a champion from the start, and Janet Kimball kept the ball rolling. Lorraine Anderson, superb copy editor, brought her sharp eyes to the pages. The entire Skipstone/Mountaineers Books team was a pleasure to work with.

Artist Biographies

Please note that page numbers where you can find the artist's work appear in parentheses after each name.

Alvin Amason (Sugpaiq/Alutiiq) was born and raised on Kodiak Island. A trailblazer in bridging traditional and modern art, and in incorporating cultural traditions, he was Director of Native Arts at UAF and developed the Alaska Native Arts curriculum at UAA. His many awards include the Alaska Governor's Award for Outstanding Visual Artist and an Honorary Degree from UAA. His work is in permanent collections worldwide. (frontispiece and pp. 48, 92, 135, and 250)

Karl Becker is a watercolor artist who lives and works in Cordova. His subject matter includes landscapes and birds. He's also inspired by the buildings in the Kennecott Mines National Historic Monument, as he enjoys depicting historic relics in their natural surroundings. He continues to be captivated by glacial landscapes and the habits of the birds he paints. (pp. 132–33 and 292)

Alison O. Bremner is a Tlingit artist born and raised in Southeast Alaska. Bremner is believed to be the first Tlingit woman to carve and raise a totem pole. She has studied under master artists David R. Boxley and David A. Boxley in Kingston, Washington. Painting, woodcarving, regalia, and digital collage are a few of the mediums the artist employs. In addition to her contemporary art practice, Bremner is committed to the revitalization of the Tlingit language and creating works for traditional and ceremonial use. Learn more at alisonobremner.com. (pp. 38, 104, 124, 184, and 227)

Courtenay Birdsall Clifford, originally from Anchorage, now makes her home in Skagway. She has followed in her parents' artistic footsteps and works primarily in a blend of acrylics and watercolors. An avid hiker, she finds unending inspiration in the Alaskan landscape and its many inhabitants, and seeks to capture the lush textures and vivid colors of nature. (pp. 6, 138, 181, 202, 232, and 270)

Sharlene Cline captures the spirit of nature in every brushstroke. She spent three years training in Taiwan under bird-and-flower brush master Yang O-Shi before making Alaska her creative home. Her work is featured in museums, galleries, and private collections across the US. (pp. 29, 112, 156, 166, and 307)

Born in Detroit in 1975, **Kevin ("KC") Crowley** moved to Alaska in 1999 to teach in the village of Nunam Iqua on the Yukon River. He later moved to Asia and started a family before returning to Anchorage where he now lives. Completely self-taught, KC began as a tattoo artist but now works primarily with woodblocks. (cover and pp. 18, 62, 108, 162, 216, and 260)

Valisa Higman's cut-paper artwork is inspired by a deep connection to both her natural surroundings and her community. She lives and works in Seldovia, and commutes to her studio by rowboat, making friends with the otters and seeing something new and interesting every day. (pp. 26, 75, 87, 150, 196, and 254)

"Chromatic" Crystal Jackson is an Iñupiaq artist from Anchorage who now creates on the Áak'w KwÁan homeland (Juneau). Since 2014, she's worked with alcohol inks in a fluid painting style. Her art is featured at the Alaska Native Medical Center, in children's books, and in private collections worldwide. (pp. 59, 77, 127, 153, 223, and 263)

Yumi Kawaguchi is a woodcut printmaker, originally from Niigata, Japan, and has lived in Fairbanks since 2001. Her inspirations come from nature and wildlife in the far north through the years of experience working in the field as well as camping trips with her husband and their Alaskan huskies. (pp. 22, 188, 194, 297, and 301)

Cheryl Lacy is an Alutiiq artist living in Wasilla. The artwork she creates reflects her journey to learn more about herself and her environment, and to introduce people to her Alaska Native culture. To be able to tell stories through artwork, whether it's from her imagination, fun things to do, or places she's been, brings her great joy. (pp. 106, 115, 208, 258, and 290)

Kristin Link is a fine artist, science illustrator, and educator who lives on the Nizina River, Ahtna Land on the edge of the Wrangell–St. Elias National Park and Preserve. She makes mixed-media work and drawings that reflect on our connection to the natural world and inspire curiosity and learning about nature. Learn more at kristinlink.com/fineart. (pp. 32, 96, 118, 141, 178, and 206)

Klara Maisch lives in Fairbanks and travels to remote regions of Alaska to paint on location. Klara is interested in many ways of learning from the land. Her artwork often highlights processes of change in the North, and she regularly collaborates with researchers to create art informed by climate science. (pp. 51, 100, 144, 236, and 294)

Kim McNett uses nature art and journaling to enhance personal connection to wild places, both at her home in Homer and on extensive wilderness travels across Alaska. She teaches visual art to all ages to encourage understanding and affection for ecosystems. (pp. 56, 69, 80, 186, 283, and 304)

Jennifer Moss, who lives and works in Fairbanks, uses contemporary visual elements to interpret the natural environment. Inspired by monochromatic winters and bright summer days, the movements of plants and animals, wonder and discovery, Moss uses a range of mediums to explore the elusive connections within our northern ecosystem. Learn more at jmossart.com. (pp. 148, 173, 215, 246, 252, and 287)

Holly Nordlum is an Iñupiaq artist, filmmaker, and activist from Alaska. She specializes in printmaking, graphic design, and traditional tattooing, working to revitalize Indigenous cultural practices. Through her art and advocacy, Nordlum raises awareness of Native identity, history, and resilience, fostering cultural pride and representation in contemporary media. (pp. 83, 158, 220, 272, and 278)

Gail Priday is a visual artist living in Fairbanks. She holds an MFA in painting and printmaking from the University of Alaska Fairbanks. Priday finds inspiration from the natural world, specifically the boreal forest. She values interdisciplinary collaborations, long walks in the woods, and spending time with her family. Learn more at gailpriday.com. (pp. 170, 175, 225, 239, 281, and 299)

Leila Pyle is an Alaskan artist who grew up in Kodiak and now lives in Fairbanks. She holds a BA in art from Reed College. Leila's work focuses on sense of place, and celebrates the ways our lives and communities are interconnected with the ecosystems we are a part of. (pp. 40, 72, 89, 129, 199, and 211)

Sara Tabbert is a printmaker and mixed media artist from Fairbanks. She holds an MFA in printmaking, and her love of woodblock printing has led to the creation of carved, painted wooden panels as well as other explorations in wood. (pp. 35, 191, 229, 267, and 274)

Ray Troll draws and paints natural-history-inspired imagery that migrates into traveling museum exhibits, books, and magazines and onto a popular line of T-shirts sold around the globe. Basing his offbeat art on the latest scientific discoveries, Ray brings a street-smart sensibility to the worlds of paleontology and ichthyology. He and his wife, Michelle, run a web store out of an ex-bordello on a salmon spawning stream in Ketchikan. (pp. 44–45, 66, 122, 234, 242, and 336)

Kesler Woodward has painted the forests, mountains, and light of Alaska and the circumpolar north for almost fifty years. In 2004 he was the first recipient of the Alaska Governor's Award for Lifetime Achievement in the Arts, and in 2012 he was awarded the Rasumson Foundation's Distinguished Artist Fellowship. (p. 14–15)

Writer Biographies

Please note that page numbers where you can find the contributor's work appear in parentheses after each name.

Martha Amore teaches writing at the University of Alaska Anchorage. She is a contributing editor of the Lambda Literary Award nominee *Building Fires in the Snow: A Collection of Alaska LGBTQ Short Fiction and Poetry*. Her short fiction collection *In the Quiet Season and Other Stories* came out in 2018. Learn more at marthaamore.com. (p. 176)

Shehla Anjum is a Pakistani American writer and longtime Anchorage resident. She works on her family's newsletters, and her writing has appeared in *Alpinist, Creative Nonfiction, First Alaskans, Anchorage Daily News*, and *Dawn* (Pakistani newspaper). Shehla has a BA from the University of Minnesota, an MPA from the Harvard Kennedy School, and an MFA from the University of Alaska Anchorage. (p. 139)

Thomas R. Bacon lives in Sitka, a small island community in the Tongass Forest. He is a member of the Blue Canoe Writers. His work has appeared in *San Pedro River Review, Borrowed Solace, Tidal Echoes, Cirque Journal, About Place Journal*, and *The Tiger Moth Review*. (p. 123)

Ray Ball, PhD, is professor of history and dean of the Honors College at the University of Alaska Anchorage. Her most recent book is the poetry collection *Trinities* (2023). She has received multiple Pushcart nominations and been a Best of the Net finalist. (pp. 90 and 273)

Chaun Ballard resides on the traditional homelands of the Dena'ina. His chapbook, *Flight*, published by Tupelo Press, was the winner of the 2018 Sunken Garden Poetry Prize. His poems have appeared in *Narrative* magazine, *Oxford Poetry, Poetry Northwest*, and the *New York Times*, and on *The Slowdown*. (p. 203)

Tara Ballard was born and raised on the traditional homelands of the Dena'ina. Author of *House of the Night Watch*, she has been published in the *Bellevue Literary Review, North American Review*, and elsewhere. She is an affiliate editor with *Alaska Quarterly Review* and assistant poetry editor at *Prairie Schooner*. (p. 146)

Toya Brown serves on the executive board of 49 Writers and is an active member of the nonprofit Sankofa. She hosts poetry events like Live from Storyknife, workshops, and panel discussions. Her work can be found in *Alaska Women Speak* and *Forum* magazine, and she has been interviewed by the Smithsonian National Museum of African American History and Culture. (p. 268)

Mike Burwell is a poet whose work has appeared in *Abiko Quarterly, Alaska Quarterly Review, Ice-Floe, Pacific Review*, and more. His poetry collection *Cartography of Water* was published in 2007, and in 2009 he founded the literary journal *Cirque*. He moved from Alaska to Taos, New Mexico, in 2013. (p. 293)

Christine Byl is the author of *Lookout*, shortlisted for the Center for Fiction's First Novel Prize as well as recipient of a Montana Book Award Honor; and *Dirt Work: An Education in the Woods*, finalist for a Willa Award in nonfiction. She is a professional trailbuilder living in Interior Alaska on the homelands of the Dene. (pp. 174 and 290)

Jerah Chadwick (1956–2016) served as Alaska's writer laureate from 2004 to 2006. He lived for seventeen years on the island of Unalaska, where he originally went to raise goats and write. He later directed the University of Alaska's extension program for the Aleutian region. He authored three chapbooks and the poetry collection *Story Hunger*. (p. 84)

Kersten Christianson is a poet and English teacher from Sitka. She is the author of *The Ordering of Stars* (Sheila-Na-Gig, 2025) and *Something Yet to Be Named* (Kelsay Books, 2017). She serves as poetry editor of *Alaska Women Speak*. Kersten savors road trips, bookstores, and smooth ink pens. (p. 35)

Anne Coray is the author of three full-length poetry collections and a chapbook on climate change. Her debut novel, *Lost Mountain*, appeared in 2021. She also co-edited the anthology *Crosscurrents North*. She divides her time between the coastal town of Homer and her birthplace on remote Lake Clark. (pp. 157 and 248)

Kim Cornwall (1967–2010) grew up in British Columbia and lived as an adult in Homer and Fairbanks. Her poems were collected by editor Wendy Erd in *Of Darkness and Light*, published by the University of Alaska Press in 2019. Her poem "What Whales and Infants Know" inspired a statewide poetry project, Poems in Place, that puts poems by Alaskans on signage in Alaska's state parks. (p. 39)

Tiffany Rosamond Creed has a BA in film studies from Portland State University and an MFA in creative writing from the University of Alaska Anchorage. Her work has been published in *The Qargi Zine, Alaska Women Speak, Into the Void, Cathexis NW,* and *Cirque*. (p. 298)

tripp j crouse (they/them, Two-Spirit Ojibwe) serves as a poetry reader for *ANMLY* and *Kitchen Table Quarterly*, and has poetry published or forthcoming in *The Yellow Medicine Review*, *beestung*, *Ink & Marrow*, and elsewhere. Their poetry chapbook, *For Every Dead Buffalo*, was published by Bottlecap Press in 2024. (p. 182)

Nora Marks Keixwnéi Dauenhauer (1927–2017) was a Tlingit poet, writer, scholar, and culture bearer from Juneau and Hoonah. She authored, with her husband Richard Dauenhauer, several acclaimed books of Tlingit oratory and oral stories. Her other books include *Life Woven with Song*, a volume of poetry and prose. The recipient of many awards, she served as Alaska's writer laureate from 2012 to 2014. (p. 253)

Bonnie Demerjian writes from Southeast Alaska's Tongass National Forest, where she migrated to teach a year or two. Before she knew it, it had become her forever home. Her poems have been published in *Alaska Women Speak*, *Tidal Echoes*, *Blue Heron Review*, *Pure Slush*, and *The Unmooring*. (p. 94)

Nancy Deschu writes nonfiction and poetry based in natural history, science, and sense of place. Her career as a field biologist and hydrologist in Alaska has provided her with a rich source of primary material and a wealth of stories. (pp. 49 and 185)

Wendy Erd's writing has been published in *Peace Works, Alaska Quarterly Review, New Rivers Press, Out on the Deep Blue,* and *Cirque*, as well as in her collection of poetry, *It's a Crooked Road But Not Far to the House of Flowers*. Her prose appears on road signs in Alaska's Copper River watershed and as poems set along an estuary trail in Homer. She directed Poems in Place, a project that placed poetry by Alaskans on signs in Alaska's state parks. (p. 251)

Daryl Farmer is the author of two books, *Bicycling Beyond the Divide* and *Where We Land*. Recent work has appeared in *Terrain.org*, *Ploughshares*, and *Natural Bridge*. He is a professor in the MFA program at the University of Alaska Fairbanks. (p. 201)

Leslie Leyland Fields lives on Kodiak and Harvester Islands, where she has worked with her family in salmon fishing for forty-five years. She's the author of fourteen books and leads writing workshops internationally as well as on both of her beloved islands. (pp. 42 and 70)

Jo Going (1948–2024) wrote poems of place and spirit in dialogue with visual art. Her book *Wild Cranes* is in the permanent collections of the Museum of Modern Art in New York and the National Museum of Women in the Arts in Washington, DC. (p. 230)

Suzi Golodoff lives on Unalaska Island and for almost fifty years has called the Aleutians home. A naturalist and writer, she's documented birdlife, published a field guide to Unalaska's wildflowers, and advocates for conservation and traditional ways of life. She also loves to write haiku. (p. 282)

Anne Hanley was Alaska's writer laureate from 2002 to 2004. She writes poetry, stage plays, and screenplays and is the co-editor of *The Alaska Reader: Voices from the North*. Since 2008, she has produced her play, *The Winter Bear*, in fifty locations all over the state. (p. 159)

Kim Heacox is best known for two of his books, *The Only Kayak*, a memoir, and *Jimmy Bluefeather*, a novel, both winners of the National Outdoor Book Award, and for his opinion pieces in *The Guardian* written in celebration and defense of the natural world. (p. 142)

Eric Heyne taught at the University of Alaska Fairbanks for thirty-seven years. He is the editor of *Desert, Garden, Mountain, Range: Literature on the American Frontier* and the University of Alaska Press edition of Jack London's *Burning Daylight*. His poetry collection is *Fish the Dead Water Hard*. (p. 212)

Robert Davis Hoffmann is a Tlingit, Tsaagweidi, of Kake. Hoffmann is inspired by his rich culture—the stories, people, history, and lands. He resides in Sitka. He has been published in numerous journals and has three books of poetry: *SoulCatcher*, *Village Boy: Poems of Cultural Identity*, and *Raven's Echo*. (p. 126)

Erin Coughlin Hollowell is a poet and writer who lives in Homer. She authored *Pause, Traveler* and *Every Atom*, poetry books published by Boreal Books. Her collection *Corvus and Crater* was published in 2023. She is the executive director of Storyknife Writers Retreat and director of the Kachemak Bay Writers' Conference. (p. 192)

Barbara Hood is a retired attorney and businesswoman who writes personal essays, commentaries, and poetry from her home in Anchorage. She is a longtime member and past board president of 49 Writers, a nonprofit organization dedicated to supporting Alaska's writing community. (p. 303)

Ishmael Khaagwáask' Hope is a Tlingit and Iñupiaq poet, Indigenous scholar, game developer, and actor who lives in Dzantik'ihéeni, Lingít Aaní. His most recent book of poetry is *Rock Piles Along the Eddy* (Ishmael Reed Publishing Company, 2017). (p. 105)

Carol Hult writes, hikes, and kayaks from her home on Kodiak Island. A teacher, parent, and writer who has lived most of her life in coastal Alaska, she has published nonfiction in *Alaska* magazine, essays in academic journals, and poetry in *Cirque: A Literary Journal for the North Pacific Rim*. (p. 107)

Joy Huntington is a Koyukon Athabascan poet and writer from the Tanana and Yukon River area. She lives in Fairbanks with her two daughters and is the principal consultant and founder of Uqaqti Consulting. Joy received a degree from Dartmouth College in Native American studies and environmental studies. (p. 167)

Laureli Ivanoff is an Iñupiaq writer from Unalakleet. Recipient of numerous journalism awards, she is a graduate of the Institute of American Indian Arts MFA program and a *High Country News* columnist. She is writing a memoir on how settler colonialism ripples into her life as an Iñupiaq woman today. (p. 78)

Nick Jans is a writer-photographer and longtime contributing editor for *Alaska* magazine. He has published more than five hundred pieces in a variety of publications, ranging from *Rolling Stone* to *USA Today*, and has also written fourteen books. His latest, *Romeo the Friendly Wolf*, is available at nickjans.com. (p. 180)

Kaylene Johnson-Sullivan lives in the foothills of the Talkeetna Mountains near the headwaters of the Susitna River. Wild places and the beings who live there have long been her inspiration and solace. Her household includes honest dogs, kindly horses, and a bemused husband. Visit her work at kaylene.us. (pp. 81 and 237)

Mar Ka is a poet whose collection *Be-hooved* (University of Alaska Press, 2019) was a finalist for the Montaigne Prize for "thought-provoking work." She lives in the foothills of the Chugach Mountains. Her chapbook *Near-Presence* won the Beyond Words 2025 Chapbook Award. (pp. 130 and 151)

Joan Naviyuk Kane, Iñupiaq from Ugiuvak and Qawairaq, is the author and editor of a dozen collections, including *Circumpolar Connections: Creative Indigenous Geographies of the Arctic, Hyperboreal,* and *with snow pouring southward past the window*. She raises her children in Oregon, where she is an associate professor at Reed College. (p. 128)

Seth Kantner was raised in northern Alaska. He's worked as a commercial fisherman, photographer, and wilderness guide, and is the award-winning author of *Ordinary Wolves* and other books, most recently A *Thousand Trails Home: Living with Caribou*. His writing and photographs have appeared in *Smithsonian, Outside,* the *New York Times*, and other publications. (p. 279)

Susheila Khera lives and works in Fairbanks. Her work has appeared in *Ice-Floe: International Poetry of the Far North*, the *Northern Review, Cirque: A Literary Journal for the North Pacific Rim,* and more. Her chapbook, *Step by Careful Step*, was published by Finishing Line Press. (p. 285)

Carolyn Kremers writes literary nonfiction and poetry from her home in Fairbanks. Her books include *Upriver*, *The Alaska Reader*, and *Place of the Pretend People: Gifts from a Yup'ik Eskimo Village*. In 2008–09 and 2015–16, she was a Fulbright Scholar at Buryat State University in Ulan Ude, Russia. Her website is pw.org/directory/writers/carolyn_kremers. (pp. 233 and 275)

Heather Lende, Alaska's writer laureate from 2021 to 2024, is the author of four memoirs from Algonquin Books: *Find the Good*; *If You Lived Here, I'd Know Your Name*; *Take Good Care of the Garden and the Dogs*; and most recently *Of Bears and Ballots* in 2020. She lives in Haines. (p. 115)

Sara Loewen is the author of *Gaining Daylight: Life on Two Islands*, winner of the Willa Award for Creative Nonfiction. She fishes for salmon with her family in Uyak Bay, on the west side of Kodiak Island. (p. 30)

Linda Martin lives in Homer. Her book of poems, *I Follow in the Dust She Raises*, was published by the University of Alaska Press in 2015. She received an MFA in creative writing from Pacific Lutheran University in 2011 and an Alaska Literary Award in 2021. (p. 133)

David McElroy is an award-winning poet who lives in Anchorage and Halibut Cove. He's enjoyed a varied work life that has included flying small planes in Arctic Alaska for forty years. His books include *Making It Simple*, *Mark Making*, *Just Between Us*, and *Water the Rocks Make*. (p. 136)

Susanna J. Mishler is the author of *Termination Dust* (Boreal Books), finalist for a Lambda Literary Award. Her poems have appeared in *Alaska Quarterly Review*, the *Iowa Review*, *Kenyon Review Online*, *Ploughshares*, and many anthologies. She is a journey-level electrician and teaches her trade to local union apprentices in Anchorage. (pp. 86 and 197)

Debbie Clarke Moderow lives in Anchorage and Denali National Park. Her memoir, *Fast Into the Night: A Woman, Her Dogs, and Their Journey North on the Iditarod Trail*, won the 2016 National Outdoor Book Award. Recent work includes poetry for a collaboration with painter Klara Maisch and scientist Rebecca Hewitt, *Threshold 32°F*. (p. 171)

John Morgan moved to Fairbanks in 1976. His work has appeared in the *New Yorker*, *Poetry*, *American Poetry Review*, and many other magazines. He won the Discovery Award of the New York Poetry Center and was the first writer-in-residence at Denali National Park. His website is johnmorganpoet.com. (pp. 209 and 265)

Vivian Yéilk' Mork (Cute-Little-Raven) is Tlingit from the Raven moiety. She was born and raised in Wrangell, but her kwáan comes from the Glacier Bay region. Yéilk' is a two-spirit ethnobotanist, traditional foods and medicines educator, writer, artist, and storyteller. She holds an MA in cross-cultural studies with an emphasis in Indigenous knowledge systems.(p. 27)

Nicole Stellon O'Donnell is the author of three books of poetry, most recently *Everything Never Comes Your Way*. She has received fellowships from the Rasmuson Foundation, an Alaska Literary Award, and a Fulbright Distinguished Award in Teaching. Her first book, *Steam Laundry*, was the Alaska Reads selection for 2018. (pp. 189 and 207)

Carrie Ayagaduk Ojanen is an Iñupiaq writer from the Ugiuvamiut tribe. She is the author of *Roughly for the North*, and her poetry has appeared in various journals. Ojanen grounds her work in familial relationships, especially with her grandparents, members of the last generation of Ugiuvak, Alaska. (p. 243)

dg nanouk okpik, Iñupiaq poet and Lannon Foundation fellow, attended Salish Kootenai College, the Institute of American Indian Arts, and Stonecoast College. Her first poetry book, *Corpse Whale*, won the American Book Award, the May Sarton Award, and the Truman Capote Award. Her 2022 collection, *Blood Snow*, was a Pulitzer finalist. (p. 60)

Jeremy Pataky is the author of *Overwinter* (University of Alaska Press, 2015). His work has appeared widely in journals. He is publisher at Porphyry Press and *Edible Alaska* and lives, works, and writes off-grid near McCarthy, within Wrangell–St. Elias National Park. (pp. 97 and 118)

Shauna Potocky is the author of *Sea Smoke, Spindrift and Other Spells* and *Yosemite Dawning: Poems of the Sierra Nevada*, both published by Cirque Press. In 2024, Shauna served as artist in residence for the Sitka Conservation Society. Shauna is deeply grateful to call Seward, her home. (p. 102)

Vivian Faith Prescott lives in Lingít Aaní in Southeast Alaska on the land of the Shtax'heen Kwáan in Wrangell. She is a member of the Pacific Sámi Searvi. She has an MFA in poetry and a PhD in cross-cultural studies, and is author of many books of poetry, fiction, and nonfiction. (p. 73)

Amy Crawford Purevsuren has worked as a wilderness guide, ski coach, Peace Corps volunteer, and public school teacher. She lives with her family in Anchorage and most enjoys exploring Alaska's wilderness with her children, by foot, kayak, bike, and skis. (p. 308)

Don Rearden is an author, screenwriter, and University of Alaska professor. He wrote *The Raven's Gift* and co-authored *Never Quit* and *Warrior's Creed*. Recent honors include a Contributions to Literacy in Alaska award and an Alaska Literary Award for fiction. *Without a Paddle* is his latest poetry collection. (p. 113)

Carol Richards is Iñupiaq and German, with family from Qikiqtaġruk. Her writing has appeared in the *Alaska Quarterly Review*, *Best Creative Nonfiction* (Volume 2), and elsewhere and earned a notable mention in *The Best American Essays* as well as a Pushcart Prize nomination. She has received residencies at Hedgebrook, Djerassi, and Storyknife. (p. 46)

Eva Saulitis (1963–2016), a marine biologist as well as poet and essayist, studied the orcas of Prince William Sound with her partner, Craig Matkin. She authored *Leaving Resurrection*, *Many Ways to Say It*, *Into Great Silence*, *Prayer in Wind*, and *Becoming Earth*. She received awards from the Rasmuson Foundation, the Alaska State Council on the Arts, and the Alaska Humanities Forum. (p. 53)

Tom Sexton (1940–2025) founded the creative writing program at the University of Alaska's Anchorage campus and taught there for many years. He served as Alaska's poet laureate from 1995 to 2000 and is the author of many poetry collections. (pp. 149 and 271)

Todd Sformo is a biologist in Utqiaġvik (Barrow) working on fish, bowhead whales, birds, and the freshwater mold *Saprolegnia*. Besides scientific papers, he has published prose poems in *Hippocampus*, *Cirque*, the *Ekphrastic Review*, and *The Fourth River* and essays in *Catamaran* and *Interalia Magazine*. (p. 295)

Bill Sherwonit, born and raised in Connecticut and an Alaska resident since 1982, is a nature writer whose work has appeared in newspapers, magazines, literary journals, and anthologies. He's the author of many books, among them *Animal Stories: Encounters with Alaska's Wildlife* and *Living with Wildness: An Alaskan Odyssey*. (p. 288)

Peggy Shumaker has been honored as Alaska's writer laureate (2010–12) and the Rasmuson Foundation's distinguished artist. Professor emerita at the University of Alaska Fairbanks, she serves on the boards of Storyknife Writers Retreat and the Raz-Shumaker Prairie Schooner Book Prizes. Shumaker is the author of eight books of poetry and a lyrical memoir. She also edits Boreal Books and the Alaska Literary Series and is a contributing editor for *Alaska Quarterly Review*. (pp. 67 and 223)

Marilyn Sigman lives in Homer. She's a lifelong naturalist, retired wildlife biologist and environmental/science educator, and former director of the Center for Alaskan Coastal Studies. Her book *Entangled: People and Ecological Change in Alaska's Kachemak Bay* received the 2020 Burroughs Medal for outstanding natural history writing. (p. 24)

Dawnell Smith finds a sense of belonging in wild places. She does communications work for Trustees for Alaska, a nonprofit environmental law firm, and lives in Anchorage with human and more-than-human beings who teach her the ways of wayfinding and relationships. She makes poems and stories when she can. (p. 187)

Frank Soos (1950–2021), writer of short stories, meditative essays, and flash nonfiction, served as Alaska's writer laureate from 2014 to 2016. He authored *Early Yet*, *Unified Field Theory* (winner of the Flannery O'Connor Award for Short Fiction), *Bamboo Fly Rod Suite*, *Double Moon* (with Margo Klass), *Unpleasantries: Considerations of Difficult Questions*, and (posthumously) *The Getting Place*. (p. 235)

Mistee St. Clair is a Rasmuson Foundation and Alaska Literary Award grantee. She is of European and Tanana Athabascan descent. Lingít Aaní (Juneau) is where she lives, hikes, writes, and wanders the mossy rain forest. Her next collection will be published by Empty Bowl Press in 2026. (p. 154)

John Straley, for many years a criminal investigator for the State of Alaska, is both a poet and the author of twelve detective novels. *The Rising and the Rain* pays poetic homage to Southeast Alaska and the Pacific Northwest, and four volumes of haiku track the seasons. Straley served as Alaska's writer laureate from 2006 to 2008. (p. 226)

Kathleen Tarr, longtime Alaskan, is the author of the memoir *We Are All Poets Here*. A Thomas Merton scholar, she frequently travels as a speaker, lecturer, and writer on Merton's literary works and spiritual legacy. Kathleen holds an MFA in literary nonfiction from the University of Pittsburgh. She lives and writes in Anchorage under the shadow of the Chugach Mountains. (p. 305)

Marie Tozier earned an MFA from the low-residency program at the University of Alaska Anchorage. She is the author of the poetry collection *Open the Dark* (2020), which illuminates elements of her Iñupiaq life in northwestern Alaska. She lives in Nome with her family. (pp. 57 and 300)

X̲'unei Lance Twitchell is a professor of Alaska Native languages, artist, musician, and author. He holds an MFA from the University of Alaska Fairbanks and a doctorate in Hawaiian and Indigenous language and culture revitalization from the University of Hawai'i at Hilo. He lives in Juneau with his family. (p. 259)

Emily Wall is a poet and University of Alaska professor of English. Her six books of poetry include the chapbooks *Fig*, *Fist*, and *Flame*, published by Minerva Rising Press. Emily lives and writes on a beach in Douglas and can be found online at emily-wall.com. (p. 227)

Erica Watson is the author of *Ghosts of Distant Trees*. She grew up in the southwestern US and now lives near Denali National Park. She earned her MFA in creative nonfiction at the University of Alaska Anchorage and has received support from Fishtrap, Storyknife, and the National Park Service. Read more at ericarobinwatson.com. (p. 195)

Annie Wenstrup is a Dena'ina poet living in Fairbanks. She's the author of *The Museum of Unnatural Histories* (Wesleyan University Press, 2025). Her poems have been published in *Alaska Quarterly Review*, *Ecotone*, *Poetry*, and elsewhere. Annie's work is supported by the CIRI Foundation, the Rasmuson Foundation, and an Alaska Literary Award. (p. 221)

Qaġġun Chelsey Zibell was raised in Noorvik. She is an Iñupiaq person and a professor of the Iñupiaq language. Her areas of study also include literature, creative writing, and education. She resides in Fairbanks with her husband, daughter, and two dogs. (p. 256)

Resources

If you'd like to learn more, here are some helpful resources—only a fraction of what's available. We encourage you to explore libraries, bookstores, and the internet. We also recommend visits to Alaska's museums and cultural centers, especially the Anchorage Museum at Rasmuson Center, the Alaska Native Heritage Center in Anchorage, the Fairbanks Museum of the North, the Alaska State Museum in Juneau, the Pratt Museum in Homer, the Sheldon Jackson Museum in Sitka, the Kodiak History Museum and the Alutiiq Museum in Kodiak, the Inupiat Heritage Center in Utqiagvik, the Museum of the Aleutians in Unalaska, and the Totem Heritage Center in Ketchikan.

FIELD GUIDES

Field guides are excellent companions in the outdoors and as references at home. In addition to the following books, many apps can be downloaded to your phone, including Merlin Bird ID, eBird, Audubon Bird Guide, PlantSnap, Picture Mushroom, Picture Nature, Seek, Plantnet, and iNaturalist.

Birds

Armstrong, Robert H. *Guide to the Birds of Alaska,* 6th ed. (Alaska Northwest Books, 2015) A great book for beginning birders, as it includes only Alaska birds and thus narrows identification possibilities. Many editions exist, each better than the previous one.

Dunn, Jon L., and Jonathan Alderfer, editors. *National Geographic Field Guide to the Birds of Western North America.* (National Geographic, 2008)

Sibley, David Allen. *Sibley Birds West: Field Guide to Birds of Western North America*, 2nd ed. (Knopf, 2016)

Plants

Golodoff, Suzi. *Wildflowers of Unalaska Island: A Guide to the Flowering Plants of an Aleutian Island*, 2nd ed. (University of Alaska Press, 2014)

Pojar, Jim, and Andy MacKinnon, editors. *Plants of the Pacific Northwest Coast: Washington, Oregon, British Columbia & Alaska*. (Lone Pine International, 2016)

Pratt, Verna E. *Field Guide to Alaskan Wildflowers Commonly Seen Along the Highways and Byways*. (Alaskakrafts, 1989) Pratt was the first to organize by color, helping amateurs with identifications.

Schofield, Janice. *Alaska's Wild Plants, Revised Edition: A Guide to Alaska's Edible and Healthful Harvest*. (Alaska Northwest Books, 2020) An abbreviated, pocket-size version of Schofield's larger volume, arranged by habitat.

———. *Discovering Wild Plants: Alaska, Western Canada, the Northwest.* (Alaska Northwest Books, 2003) An encyclopedic volume with photos, illustrations, historic and cultural information, and considerable information about food and medicinal use.

Studebaker, Stacy. *Wildflowers and Other Plant Life of the Kodiak Archipelago: A Field Guide for the Flora of Kodiak and Southcentral Alaska.* (Sense of Place Press, 2010)

Viereck, Leslie A., and Elbert L. Little, Jr. *Alaska Trees and Shrubs*, 2nd ed. (University of Alaska Press, 2007)

White, Helen A., and Maxcine Williams, editors. *Alaska-Yukon Wild Flowers Guide*. (Alaska Northwest Books, 1974)

Marine

Barr, Lou and Nancy. *Under Alaskan Seas: The Shallow Water Marine Invertebrates.* (Alaska Northwest Books, 1983)

O'Clair, Rita M., and Sandra C. Lindstrom. *North Pacific Seaweeds.* (Plant Press, 2000) Includes a wealth of information, illustrations, photographs, and even recipes.

Wynne, Kate. *Guide to Marine Mammals of Alaska*, 4th ed. (Alaska Sea Grant College Program, 2013) A spiral-bound book handy for boat travel, with photos, illustrations, range maps, and plenty of information.

Regional

Carstensen, Richard, Robert H. Armstrong, and Rita M O'Clair. *The Nature of Southeast Alaska: A Guide to Plants, Animals, and Habitats*, 3rd ed. (Alaska Northwest Books, 2014)

O'Harra, Daniel, Katherine Hocker, Kristan Hutchison, et al. *Alaska's Kenai Peninsula Wildlife Viewing Trail Guide.* (Alaska Department of Fish and Game, 2007) Although specific to the Kenai Peninsula and sites there, includes viewing tips and hints, field and habitat notes, and information about cultural, historic, and geological connections.

Pielou, E. C. *A Naturalist's Guide to the Arctic.* (University of Chicago Press, 1995)

Other Field Guides

Connor, Cathy. *Roadside Geology of Alaska*, 2nd ed. (Mountain Press, 2014)

Dilley, Lorie M., and Thomas E. Dilley. *Guidebook to Geology of Anchorage, Alaska*. (Publication Consultants, 2001)

Laursen, Gary A., and Neil McArthur. *Alaska's Mushrooms: A Wide-Ranging Guide*. (Alaska Northwest Books, 2016)

Leahy, Christopher. *Peterson First Guide to Insects of North America*. (Mariner Books, 1998) What Alaska has in insect biomass (mosquitoes!) it lacks in numbers of species, but this book can be a great start for, among other things, telling the difference between a no-see-um and a black fly.

Murie, Olaus J. *A Field Guide to Animal Tracks*, 2nd ed. (Houghton Mifflin, 1975) The complete guide, with a wealth of wildlife information and illustrations of the animals, their tracks, and their droppings.

Pandell, Karen, and Chris Stall. *Animal Tracks of the Pacific Northwest*. (Mountaineers Books, 1981) A pocket-size guide that includes field notes as well as illustrations of tracks.

Parker, Harriette. *Alaska's Mushrooms: A Practical Guide*. (Alaska Northwest Books, 1994) Very practical—where and when to find, features, look-alikes, food uses and cautions.

Philip, Kenelm W., and Clifford D. Ferris. *Butterflies of Alaska*. (Self-published, 2015)

ALASKA NATIVES AND TRADITIONAL KNOWLEDGE

Worldviews / Cosmology

Dauenhauer, Nora Marks, and Richard Dauenhauer. *Haa Kusteeyí, Our Culture: Tlingit Life Stories.* (University of Washington Press, 1994)

Fienup-Riordan, Ann. *Nunakun-gguq Ciutengqertut/They Say They Have Ears Through the Ground: Animal Essays from Southwest Alaska.* (University of Alaska Press, 2020)

Kalifornsky, Peter. *A Dena'ina Legacy: K'tl'egh'I Sukdu: The Collected Writings of Peter Kalifornsky.* (Alaska Native Language Center, 1991)

Kawagley, Angayuqaq Oscar. *A Yupiaq Worldview: A Pathway to Ecology and Spirit*, 2nd ed. (Waveland Press, 2006)

Luke, Howard. *Howard Luke: My Own Trail.* (Alaska Native Knowledge Network, 1998)

Napoleon, Harold. *Yuuyaraq: The Way of the Human Being*, 5th ed. (Alaska Native Knowledge Network, 1996)

Worl, Rosita Kaahani. *Native Values: Living in Harmony.* (Sealaska Heritage Institute, 2017)

Plants

Fienup-Riordan, Ann, Alice Rearden, Marie Meade, Kevin Jernigan, Jacqueline Cleveland, Sharon Birzer, Richard W. Tyler. *Yungcautnguuq Nunam Qainga Tamarmi/All the Land's Surface Is Medicine: Edible and Medicinal Plants of Southwest Alaska* (University of Alaska Press, 2021)

Kari, Priscilla Russell. *Tanaina Plantlore: An Ethnobotany of the Dena'ina Indians of Southcentral Alaska* (Alaska Native Language Center, 1995)

Art

Fair, Susan W. *Alaska Native Art: Tradition, Innovation, Continuity.* (University of Alaska Press, 2007)

Traditional Knowledge Online

Alaska Native Knowledge Network, www.uaf.edu/ankn.

Indigenous Knowledge and Traditional Ecological Knowledge, Alaska (National Park Service), nps.gov/subjects/tek/alaska.htm.

"Indigenous Knowledge Systems and Alaska Native Ways of Knowing" by Ray Barnhardt and Angayuqaq Oscar Kawagley, www.ankn.uaf.edu/curriculum/Articles/Barnhardt Kawagley/Indigenous_Knowledge.html.

Yup'ik Bird Book, Alaska Native Knowledge Network, www.uaf.edu/ankn/alaska-native-cultural-re/alaska-native/yupik-bird-book/.

Language Dictionaries

Language revitalization is a growing movement in Alaska, and it's fun and informative to learn even a few names in the places where they live. A number of language dictionaries are now online or can be downloaded in pdf form, including the following:

Ahtna (place names only), uafanlc.alaska.edu/Online/AT973K1983b/AT973K1983b.pdf.
Dena'ina, uafanlc.alaska.edu/Online/TI974WK1979/wassillie-1979-den_dictionary.pdf.
Gwich'in, uafanlc.alaska.edu/Online/KU973P1979a/ga26.pdf.

Iñupiaq, inupiaqonline.com/.
Lower Tanana Athabascan/Minto, uafanlc.alaska.edu/Online/TN990T2015a/TN990T2015a.pdf.
Tlingit, Haida, and Tsimshian, sealaskaheritage.org/shi-tlingit-haida-and-tsimshian-audio-dictionary/.
Yup'ik, www.swrsd.org/site/handlers/filedownload.ashx?moduleinstanceid=254&-dataid=274&FileName=Yupik_Eskimo_Dictionary_Vol_1.pdf.

Permissions

Every effort was made to reach the copyright holders of the works included in this volume. Contributing poets, writers, and artists graciously allowed the use of their work. Additional notes and permissions include the following:

"Field Guide to the Ghost Forest" and "You Say the Mosquitos Are Terrible This Year" by Ray Ball. First published in *Alaska Women Speak* (Summer 2021). Reprinted by permission of the author.

"A Poem Ending with a Strambotto Wherein I Include an Extra Line That Is Myself *or* A Poem in Which I Name the Flower" by Chaun Ballard. First published in the *Missouri Review*, 2024, and collected in *Second Nature*, BOA Editions, 2025. Reprinted by permission of the author.

"Magic Maybe" by Tara Ballard. First published in the *Missouri Review*, 2024. Reprinted by permission of the author.

"Curry Ridge" by Mike Burwell. First published in *Cartography of Water* (Northshore Press, 2009). Reprinted by permission of the author.

"Lynx" by Christine Byl excerpted from *Dirt Work: An Education in the Woods* (Beacon Press, 2013). Reprinted by permission of the author and publisher; permission conveyed through Copyright Clearance Center, Inc.

"From the Museum" by Jerah Chadwick. First published in *Story Hunger* (Salmon Poetry, 1999). Reprinted by permission of the publisher.

"Carol's Laminaria" by Kersten Christianson. First published in *Something Yet to Be Named* (Kelsay Books, 2017). Reprinted by permission of the author.

"*Heracleum lanatum* (Cow Parsnip)" by Anne Coray. First published in *Bone Strings* (Scarlet Tanager Books, 2005) and reprinted by permission of the author and publisher. "*Sterna paradisea*" by Anne Coray. First published in *A Measure's Hush* (Boreal Books, 2011) and reprinted by permission of the author and publisher.

"What Whales and Infants Know" by Kim Cornwall. First published in *Of Darkness and Light* (University of Alaska Press, 2019). Reprinted by permission of the University Press of Colorado.

"Bagone'giizhig (Hole in the Sky)" by tripp j crouse. First published in *For Every Dead Buffalo* (Bottlecap Press chapbook). Reprinted by permission of the author.

"Spring" by Nora Marks Keixwnéi Dauenhauer. First published in *Raven Tells Stories* (Greenfield Review Press, 1991). Reprinted by permission from executor Carmela Ransom.

"See, Lions" by Bonnie Demerjian. First published in *Alaska Women Speak* (Spring 2024). Reprinted by permission of the author.

"A Form of Prayer" by Wendy Erd. First published in *Cirque* and reprinted in *It's a Crooked Road but Not Far to the House of Flowers* (Poetry Box, 2023). Reprinted by permission of the author.

"The Halibut" by Leslie Leyland Fields. First published in *The Water Under Fish* (Trout Creek Press, 1995). Reprinted by permission of the author. "Tideline" by Leslie Leyland

Fields. First published as part of the Poems in Place project as a placard in Fort Abercrombie Historical Park, Kodiak. Reprinted by permission of the author.

"Death and the Earth's Weather" by Eric Heyne. First published in *Fish the Dead Water Hard* (Cirque Press, 2021). Reprinted by permission of the author.

"Raven Moves" by Robert Davis Hoffman. First published in *Raven's Echo* (University of Arizona Press, 2022). Reprinted by permission of the publisher.

"skating after many moons" by Marybeth Holleman. First published in *The Hopper* (2020) and collected in *tender gravity* (Boreal Books, 2022). Reprinted by permission of the author and publisher. "Sundew: from that which appears inconsequential" by Marybeth Holleman. First published in *Entropy* (2019) and collected in *tender gravity* (Boreal Books, 2022). Reprinted by permission of the author and publisher.

"Choir hive" by Erin Coughlin Hollowell. First published in *Orion* (2020) and collected in *Corvus and Crater* (Salmon Poetry, 2023). Reprinted by permission of the author and publisher.

"Making *Achaaqhluk*" by Laureli Ivanoff. Excerpted from "How cooking, eating and harvesting beach greens ties a family together," published in *High Country News*, May 27, 2022. Used by permission of the author and publisher.

"The Circle of the Kill" by Nick Jans. Adapted from "The Circle of the Kill" in *The Last Light Breaking* (Alaska Northwest Books, 1993).Used by permission of the author.

"Eclipse of Autumnal Moths" by Mar Ka. First published in *Be-hooved* (University of Alaska Press, 2019). Reprinted by permission of the author and the University Press of Colorado.

"Hyperboreal" by Joan Naviyuk Kane. First published in *Hyperboreal* (University of Pittsburgh Press, 2013). Reprinted by permission of the publisher.

"Harboring a Mean Streak" by Linda Martin. First published in *I Follow in the Dust She Raises* (University of Alaska Press, 2015). Reprinted by permission of the author and the University Press of Colorado.

"Bear 747" by David McElroy. First published in *Water the Rocks Make* (University of Alaska Press, 2022). Reprinted by permission of the author and the University Press of Colorado.

"Brittle Stars" and "Throw Salt" by Susanna J. Mishler. First published in *Termination Dust* (Boreal Books, 2014). Reprinted by permission of the author and publisher.

"November Surprise" and "Counting Caribou—Prudhoe Bay, Alaska" by John Morgan. Both first published in *Archives of the Air* (Salmon Poetry, 2015). Reprinted by permission of the author and publisher.

"Lonely Owl" and "Picking Crowberries" by Nicole Stellon O'Donnell. First published in *Everything Never Comes Your Way* (Boreal Books, 2021). Reprinted by permission of the author and publisher.

"the soul is a" by Carrie Ayagaduk Ojanen. First published in *Roughly for the North* (University of Alaska Press, 2018). Reprinted by permission of the author and the University Press of Colorado.

"Whiteout Polar Bears" by dg nanouk okpik. First published in *Blood Snow* (Wave Books, 2022). Reprinted by permission of the author and publisher.

"The Ice of Norton Sound, Refusing" by Eva Saulitis. First published in *Many Ways to Say It* (Red Hen Press, 2012). Reprinted by permission of the publisher.

"Magpie at Twilight" by Tom Sexton. First published in *A Ladder of Cranes* (University of Alaska Press, 2015). "Snowy Owl" by Tom Sexton. First published in *For the Sake of the Light* (University of Alaska Press, 2009). Reprinted by permission of the author and the University Press of Colorado.

"Swans, Where We Don't Expect Them" by Peggy Shumaker. First published in *Bellingham Review*. "Rapt" by Peggy Shumaker. First published in *Cream City Review*. Reprinted by permission of the author.

"The Blue Fish" by Frank Soos. First published as part of the Poems in Place project (2013) and displayed at the Chena River State Recreational Area. Used by permission of Margo Klass.

"Becoming Moss" by Mistee St. Clair. First published in *Tidal Echoes*, 2024. Reprinted by permission of the author.

Two untitled haiku by John Straley. First published in *100 Poems of Fall* (Shorefast Editions, 2017). Reprinted by permission of the author.

"King Crab" and "Lingonberries" by Marie Tozier. Both first published in *Open the Dark* (Boreal Books, 2020). Reprinted by permission of the author and publisher.

"Ggugguyni Translates the Archive as Diviner" by Annie Wenstrup. First published in *Green Linden Press*, Issue 14, 2022, and collected in *The Museum of Unnatural Histories* (Wesleyan University Press, 2025). Reprinted by permission of the author and Wesleyan University Press.

Beings Index

Art and Artists Index

Entries are listed in order of appearance.

About the Editors

Marybeth Holleman was raised by North Carolina's Great Smoky Mountains and lives in the embrace of Alaska's Chugach Mountains. She's the author of the novel *Bloom Again*, the poetry collection *tender gravity*, and the memoir *The Heart of the Sound*, as well as co-author of *Among Wolves* and co-editor of *Crosscurrents North*, among others. Her award-winning work appears in more than fifty publications, including *Orion*, *Christian Science Monitor*, *Sierra*, *North American Review*, *zoomorphic*, and *The Guardian*. She's held artist residencies in such diverse places as Hedgebrook, Mesa Refuge, Ninfa, Denali National Park, and Tracy Arm-Fords Terror Wilderness. She transplanted to Alaska after falling head over heels for Prince William Sound just two years before the oil spill. When she's not kayaking those beloved fjords, she's following her wild huskies up and down Alaska's mountains. www.marybethholleman.com (pp. 214, 240)

Nancy Lord, a former Alaska state writer laureate (2008–2010), is the author of three short story collections, five books of literary nonfiction including *Beluga Days: Tracking a White Whale's Truths* and *Early Warming: Crisis and Response in the Climate-Changed North*, and the 2017 novel *pH*. She also edited the anthology *Made of Salmon*. Her work, which focuses mainly on environmental and marine issues, has appeared widely in journals and anthologies and has been honored with fellowships and awards. She holds degrees from Hampshire College (liberal arts) and Vermont College of Fine Arts (MFA in fiction writing). She taught creative writing in the University of Alaska system for many years and since 2015 has taught science and nature writing for Johns Hopkins University. She reviews Alaska-related books for the *Anchorage Daily News*, serves as an advisor to Storyknife Writers Retreat and the Kachemak Bay Writers' Conference, and frequently adjudicates writing contests. www.writernancylord.com (p. 33)

Shaelene Grace Moler (she/her) grew up in K̲éex̲' K̲waan, the community of Kake. Her Tlingit name is Sgweín and she is of the Tsaagweidí, the split-finned killer whale clan, from the yak's lits'eix̲i hít, the house that once anchored the village. She graduated with a double major in English and environmental studies with an emphasis in creative writing from the University of Alaska Southeast. She has been published in *First Alaskans* magazine, *Juneau Empire*, *Tidal Echoes*, and *Alaska Women Speak* and has done writing fellowships with *First Alaskans* magazine and Alaska Humanities' *Forum* magazine. She also served as a lead editor for the literary and art journal *Tidal Echoes*. (p. 76)

Ray Troll 2017

ABOUT SKIPSTONE

Skipstone is an imprint of independent, nonprofit publisher Mountaineers Books. It features thematically related titles that promote a deeper connection to our natural world through sustainable practice and backyard activism. Our readers live smart, play well, and typically engage with the community around them. Skipstone guides explore healthy lifestyles and how an outdoor life relates to the well-being of our planet, as well as of our own neighborhoods. Sustainable foods and gardens; healthful living; realistic and doable conservation at home; modern aspirations for community—Skipstone tries to address such topics in ways that emphasize active living, local and grassroots practices, and a small footprint.

Our hope is that Skipstone books will inspire you to effect change without losing your sense of humor, to celebrate the freedom and generosity of a life outdoors, and to move forward with gentle leaps or breathtaking bounds.

All of our publications, as part of our 501(c)(3) nonprofit program, are made possible through the generosity of donors and through sales of 700 titles on outdoor recreation, sustainable lifestyle, and conservation. To donate, purchase books, or learn more, visit us online:

www.skipstonebooks.org
www.mountaineersbooks.org